OTT Business Opportunities

By Lawrence Harte

OTTBusiness.com

DiscoverNet Publishing
1105 Walnut Street, Suite F160-25
Cary, NC 27521 USA
Telephone: +1.919.301.0109
email: info@DiscoverNet.com
web: www.DiscoverNet.com

DiscoverNet SOLUTION MARKETING

International Standard Book Number: 9781932813944

About the Author

Hi. I'm Lawrence Harte, creator of this book and its companion course. My wide array of experiences in the field of TV technology and business inspired me to teach others discover (new) ways to implement OTT systems and services. My mission has been to explain OTT options and provide resources to help you set up, run and optimize OTT systems and services.

I am the founder and senior editor for several TV tech & business magazines, including OTT Business Magazine, IPTV Magazine, Social TV Magazine, Blockchain Media Magazine and other publications. I am also the podcast host for Internet TV Plus and Blockchain Media.

As a TV industry journalist and book author since 1991, I've interviewed many successful OTT executives and experts as well as over 4,100 companies since 2005. During my visits to trade shows, including NAB, IBC and CES, I typically interview 70 to 100 companies at each event.

I'm the founder and CEO of DiscoverNet where I manage a school, run trade shows, and publish magazines and books. Running and growing these business units allows me to understand and feel some of the pains and challenges you are experiencing.

I am also the CTO of Ski TV, where I discover, recommend and set up new TV systems, services and applications and the CMO of VegiPlus.tv, an MTV-like healthy living TV network and media platform.

Some of the information presented is used in my client's OTT systems to help them create new Streaming TV services and earn additional revenue(s), both on and off screen. Some benchmarks and processes covered in the book come from my clients' own experiences.

Over the years, I've worked for major companies including Google TV, Samsung, Ericsson and Rovi and over 20 others.

Expert Contributors

Chris Wagner
OTT Services

Chris Wagner is a technology expert who partners with investors, executives, and entrepreneurs to grow their professional brands, revenues, and customers. After spending more than a decade in the video internet streaming industry, Chris has developed an uncanny ability to start-up new technology businesses and enhance existing enterprises through digitally enabled services. He co-founded internet start-up NeuLion and helped grow the business to $100 million dollars, which sold to Endeavor for $250 million in cash. Chris's ability to creatively apply technology to business opportunities has given him a front row seat with management teams, boards, and the internet industry. His passion for technology and how to apply it to enable business success, has created a network of followers interested in his views and how he might add value to their teams.

Tim Bell
Video Production

Tim Bell is president of Blackett Bell Productions and is a film and TV show producer who specializes in developing creative concepts and developing effective content. Tim has created, produced, and directed. His projects range from episodic content to full feature films. Tim specializes in creating OTT specific content to increase reach and improve engagement with audiences along with new revenue opportunities. He is dedicated to being a part of productions that can achieve monetary goals and is passionate about creating content that connects with people in mind and soul.

Michael Nagle is Founder and CEO of Ashling Digital - a consulting firm focusing on streaming TV distribution and marketing for studios, content owners, and media networks. Michael has setup, negotiated, and managed 280+ OTT and TV distribution agreements. He serves as Head of Distribution for Ski TV, DocuBay and RISE Revolutions Cycling Network. He's managed global distribution and marketing for a wide array of brands including The Box Music Network, Bloomberg Television, Playboy, and NatureVision TV. At Bloomberg, Michael represented the company during the FCC investigation surrounding independent programming, Broadcast Must-Carry and also a la carte packaging regulation. He was named one of Home Media Magazine's "Digital Drivers in Media" in both 2016 & 2017 for his efforts on behalf of Invincible Entertainment for whom he serves on their Advisory Board.

Gladston Gomez
OTT Networks

Gladston Gomez is a OTT/IPTV Consultant / Technology Expert supporting various Media Broadcasting and Telecom Companies in Design, integrate and create various workflows like Linear Live Streaming , VOD , NPVR , CATCHUP , NETWORK TIMESHIFT , Adtech Etc . Built Business case for OTT, SaaS TV solution, Cloud Based TV Solution, WebTV, IPTV, Hybrid-TV, TV-Middleware, CAS, DRM, CMS, CDN, BSS, TV Recommendation, TV- Analytics, TV Advertisement Management, Video Processing System . Proven track record in partner development, solution selling and building high performance teams. Ability to network with CXO's and nurture professional relationships at all levels. Provide technical guidance in building the OTT platform across platforms (i.e. Android, iOS,Connected TV, Set Top Box etc.) Training Technology team on the systems, workflows, standard operating procedures, disaster recovery, failover procedures.End to End Project management for technical setup of the channels and re-engineering.

Eve Glover
Creative Media

Eve Glover is an arts & culture/lifestyle writer who has been published in American and Israel newspapers and magazines, including The Jerusalem Post, Israel National News, The Jewish Press and The American Israel Numismatic Association's Shekel Magazine. Her writing has also appeared in lifestyle, travel and health publications such as Reader's Digest, The Healthy, Marriott International's in-room travel magazine, JWM Magazine, Hawaiian Airlines' in-flight lifestyle magazine, Hana Hou!, Aspen Peak, Boston Common, Beverly Hills Magazine, Luxury Lifestyle Magazine and JMG Lifestyle Magazine. As an editor, Eve has written and edited medical/research material and edited music/media content. She is also lyricist who has collaborated on songs with Grammy, Emmy and hit music producers.

Acknowledgements

Many smart people have helped to create this book. Some of them gave substantial amounts of time to share their experience, answer many questions, and invite us into their businesses and onto their production sets.

Book Contributors and Editors including Chris Wagner from OTT Advisors, Eve Glover - Writer & Lyricist, Gladston Gomez a Media and Broadcasting Consultant, Michael Nagle of Ashling Digital, and Timothy Bell from Blackett Bell Production.

OTT System and Service Professionals including Ben Weinberger - Investor & Entrepreneur, Bud Bates with TCIC, Chris Pfaff from Chris Pfaff Tech Media, Dan Bigman with Chief Executive Group, Ellen Feaheny from AppFusions, Gabriel Baños with Flowics, Hank Frecon at Source Digital, Ian Locke from Minerva Networks, Ian McDonough of Blackbird Video, Jeff Greenfield with C3 Metrics, Jim O'Neill and Jonathan Witte and Lexie Knauer from Brightcove, Madeleine Noland at ATSC, Mauro Bonomi with Minerva, Michael Lantz of Accedo, Michael Malcy with Decentrix, Mika Rautiainen of Valossa Labs, Per Lindgren at Net Insight, Peter Maag with Haivision, Rick Brown of NC State University, Roger Franklin from LTN Global, Roger McGarrahan of Pathfinder, Ron Van Herk at Cybermedia Television BV, Ross Cooper with Channel Islands, Sam Orton-Jay of V-Nova, Scott Davies at Never.no, Taras Bugir with Decentrix and Toni Leiponen   at Icareus.

TV and Movie Production leaders and on air talent including Billy Lewis from Orange Films, Budd Margolis with TV tuyo, Curt Marvis of QYOU Media, Drew Becker at Convey Media, Evelina Grines with Decor Steals, Francois Quereuil from Avid, Frank Timberlake of R F Timberlake & Company, George Wehmann at Direct Marketing Resource Group, Jerry McGlothlin from Moving Visions Entertainment, John Clark with ABC 11, John Daly at Undercover Jetsetter, John Demers from Rusty Bucket, Kim Brame with Creative Illusions, LaMont Johnson with Chieko Alice Films,

Melissa St. John of Loop Creative, Mike Davis at Uptone Pictures, Neil Roberts Post Production Specialist, Omar McCallop from Galaxy Studios, Ray Williams of Crumbs Music, Scott Markowitz with Contrast Creative, and Scott Rucci of Rucci Productions.

TV Broadcast and OTT Analysts and Consultants including Allan McLennan with PADEM Media Group, Colin Dixon of nScreenMedia, Rick Ducey from BIA, Simon Murray with Digital TV Research, Tom Weiss of Marketcast, and Scott O'Neill with MPP Global.

TV Association, OTT Event Hosts and Industry Media including Brian Mahony from Trender Research OTT Exec, Dan Rayburn from the NAB Show, Laura Riggs of Laura Riggs Consulting, Lucy Meek with the IABM, Rich Tehrani of TMC, Rupin Kotecha at Digital Guru, Stacey Orlick with Informa, Stan Moote at IABM and Tracy Swedlow with TV of Tomorrow (TVoT).

Special thanks to advisors and business supporters including Buddy Howard, Ben Levitan, Denis McDuff, Roman Kikta and Robert Belt.

Table of Contents

CHAPTER 6 - TELEVISVION COMMERCE (TCOMMERCE)

Chapter 1

OTT Business

This book helps OTT service providers, TV broadcasters, system & equipment providers, content owners, and others involved in the streaming TV - OTT industry - to understand opportunities, options, and get information to help implement news services.

OTT Business

- New OTT Services
- Trends & Benchmarks
- Key Terms & Acronyms
- Enhanced Advertising
- TV App Options
- Social TV Monetization
- eCommerce on TV

OTT Business Opportunities
OTT Services
OTT Advertising
TV Apps
Social TV
tCommerce

OTTBusiness.com

OTT Business Needs

OTT companies and related businesses have these key interests:

New OTT Services - Discover new ways to earn revenue, preferably quickly with little investment and reasonable risk.

Trends and Benchmarks - Learn about potential sales opportunities from results driven examples. This is helpful for planning and also useful for getting investment.

Key Terms and Acronyms - Learn how to understand and communicate with others in the field by speaking their language. You don't want to risk sounding like a novice unaware of key OTT terms.

Enhanced Advertising - Explore ways to increase ad revenue through ad targeting, and learn how to sell and manage ads on OTT and partner platforms.

TV App Options - Find out how and why to create TV and mobile apps for distribution to earn revenue.

Social TV Monetization - Discover different ways to engage audiences and make money from doing it.

eCommerce on TV - Learn practical ways to transition the $4.2 trillion eCommerce industry into the television industry without having to make any changes to devices or systems.

Key OTT Services and Systems

OTT systems provide people with online access to live or on demand programs on multiple types of connected devices. OTT provider key drivers include content services, platforms and audience engagement.

OTT business can be divided into these key areas: OTT Services, Advertising, TV Apps, Social TV and tCommerce.

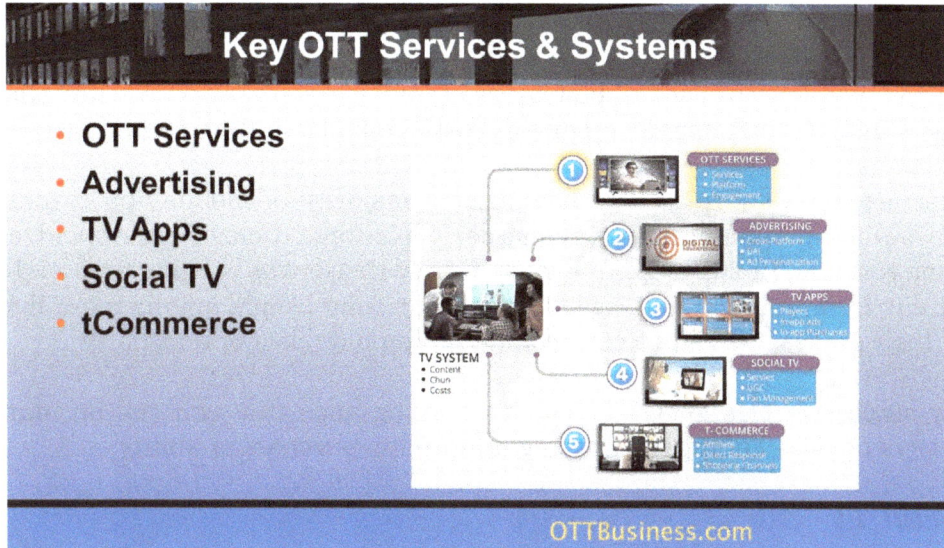

Key OTT Services & Systems

- OTT Services
- Advertising
- TV Apps
- Social TV
- tCommerce

OTTBusiness.com

OTT Service Providers (OSPs)

In 2020, there were over 600 key streaming TV providers in the USA. Several of these are highly profitable with over 10 million subscribers. Surprisingly, some that have the highest growth rates and profit margins are not NetFlix, Disney, or Hulu.

OTT providers can focus on key content topic areas such as sports, education, documentaries and other categories.

Key success factors for many OTT service providers are niche content types, subscriber growth, and user experiences (UX).

OTT Advertising

OTT advertising includes video advertising and digital ads. Because OTT advertising can be targeted and personalized, it can earn 10x or more ad revenue compared to traditional broadcast advertising.

Key success factors for OTT advertising include subscriber ad targeting, cross-channel advertising and personalized interactive ads.

OTT TV and Mobile Apps

TV and mobile Apps control how users obtain streaming services. Key challenges include high app development costs and distribution channel fees.

Most people are surprised to hear that a majority of mobile app revenue comes from in-app purchases. A smaller percentage comes from App advertising revenue. In 2015, the free game Candy Crush was earning over $1.1M per day from in-app purchases, most coming from people buying more lives for their games. OTT apps can include in-app purchase options.

Key success factors for OTT apps are having apps that can run on many types of devices, in-app advertising and in-app purchasing ability.

Social TV

Social TV services provide new ways for OTT providers to attract and engage with audiences, get user generated content (UGC), as well as organize and monetize fan engagement.

Social TV has gotten a bad reputation from second screen TV market failures. It was anticipated that viewers would simultaneously watch related TV show content on their Smartphones. While most viewers do use Smartphones while watching TV, they usually access other services and content during the shows. The creation of second screen systems and services had very low interest which resulted in big losses for second screen content and service providers.

There is enormous value for social TV in attracting viewers (sharing video clips), motivating social media viewers to become subscribers, and earning a portion of the ad revenues automatically displayed on social networks. OTT providers who share program related media on social networks such as Youtube and Facebook earn 55% of the $110B advertising spent on social networks each year.

Social TV key success factors include using video clipping and social media to attract and engage audiences, motivating user content contributions and setting up social network ad revenue sharing channels.

Television eCommerce (tCommerce)

Television Commerce (tCommerce) is the providing of eCommerce on TV systems. The global eCommerce was $4.2 Trillion in the beginning of 2020 compared to the $226 Billion global pay TV industry - subscriptions and pay per view for cable TV, satellite, and paid local TV services.

Many tCommerce services have been tried and failed since the 1970s. These systems include Qube interactive TV (1977), Full Service Network in Orlando Florida (1994), Sky Digital interactive football and OnNet eCommerce (2000) and many others. All of these systems required special television adapter boxes and changes to the broadcast TV system. Providing tCommerce on OTT systems can be done using TV apps and mobile apps which do not require special adapters or changes to TV systems.

tCommerce key success factors include setting up affiliate referral programs that earn commissions on and off screen and having interactive shopping channels and direct sales programs.

OTT Industry Players

Key players in the OTT industry include TV broadcasters and OTT providers, such as OSPs, content owners and equipment and service providers. There are also many other types of companies that are part of the OTT industry who benefit from its growth, so long as they can adapt to the changing marketplace.

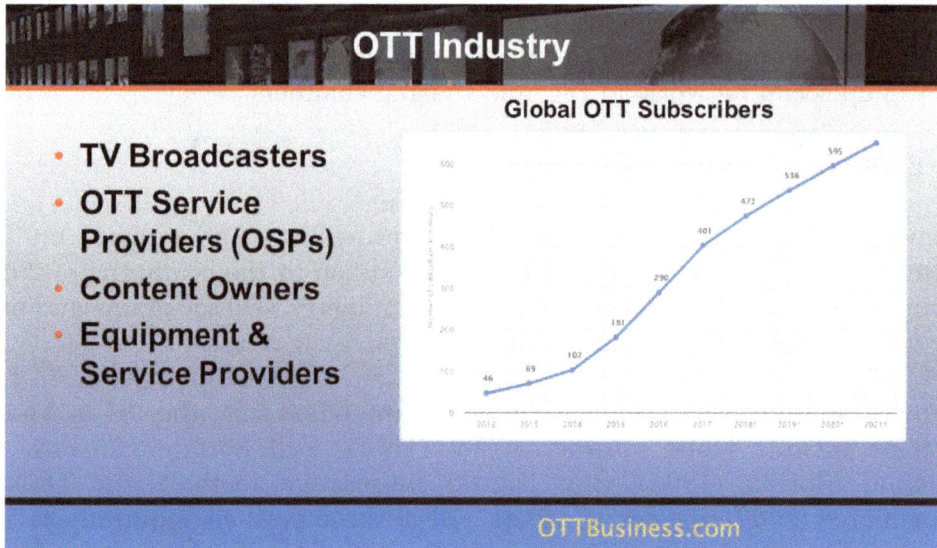

TV Broadcasters

TV broadcasters are companies that transmit or provide televised information to users that are connected or able to access signals on the broadcast
network. TV broadcasters gather and license content, process and group it
into channels, and distribute it directly to customers or through affiliate
partners. TV broadcasters can be transmitted through local wireless (terrestrial) channels, satellite, or on cable TV systems.

OTT Service Providers (OSP)

Over the top (OTT) service providers (OSPs) are companies that provide
television or video services that connect through the internet or other types
of data networks. OSPs gather and license content, process and group it into
streaming channels or on demand programs and typically distribute it
directly to customers through the internet. While it is technically possible
for OSPs to provide services globally, content licensors and regulators typically limit the geographic areas that OSPs can provide services to.

Content Owners

Content owners are companies that produce or own the rights to movies, TV shows or other media that may be provided by an OSP. Content owners may directly license their content for distribution directly to broadcasters and TV networks, or through distributors who typically transmit to smaller TV systems.

Equipment Manufacturers

Equipment manufactures create devices or assemblies that enable media companies to produce, distribute or manage programs and shows. This specialized equipment includes video editors, digital compressors (encoders), switching systems and other types of processing devices.

Cloud Services Providers

Cloud service providers set up and run online applications for media production, storage, editing and distribution equipment that are accessed through the internet. In general, the media industry is transitioning from dedicated media processing equipment to cloud services. Specialized equipment is being replaced with computer servers that can be continually updated with new protocols and applications.

OTT Resources

Valuable resources are included in this book to help you find, set up and run OTT systems.

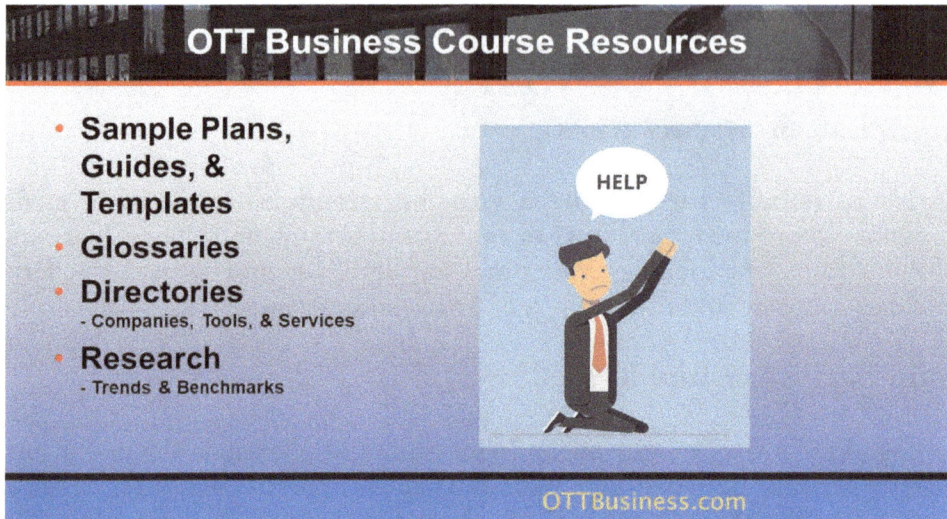

Glossaries - Terms and acronyms used in this book. These terms help you decode the language of experts. You don't want to look like an amateur when talking with industry cohorts.

Directories - Includes lists of companies, tools and services for each lesson.

Research - Has market data trends and implementation examples.

Resources - Sample plans, agreements, guides, templates and other informational documents are available.

The book is continually updated and more resources are also added. If you have a request for a book topic or resource, please contact us.

Hope you like to learn! There are most likely many topics and terms in this book you haven't heard of - yet. There are also several surprises! When you are done, you should understand and feel excited about the opportunities for the OTT industry.

Chapter 2

OTT Service Providers (OSPs)

OTT service providers (OSPs) are companies that can stream and/or download movies, TV shows and other video content through the internet.

OTT providers primarily focus on Content, Services, Platform, Workflow, Devices, and Subscribers.

What OTT companies want to know about providing streaming TV services:

OTT Service Types - What types of services should I provide - should we offer live, on demand, subscription or advertising funded? What types of service packages or bundles can I offer?

Content Types and Sources - How to determine types of content we should have? Should we have multiple types of programming like Netflix? Not likely because it's too costly. Should it be specialized content, such as sports, sci-fi or horror? Should we produce original content? Not usually because it's complicated and expensive. Where can we get our content and what license terms do they typically provide - one month access, on demand, content security requirements?

Streaming Platform Options - How do I gather, organize and stream the content? Can I get movies and shows, as well as transfer movies and shows, through the internet, storage disks and thumb drives? Will I build my own platform? Not usually. What are important platform options and how will my platform choice affect my feature options, production processes and streaming costs?

Revenues & Benchmarks - What are my revenue sources - monthly subscription fees, ad revenues, sponsorships, affiliate commissions, other sources? What are typical revenues and how do my OTT service revenues and margins compare?

Viewing Devices - What types of devices will my system work with - Smart TVs, digital media adapters - media sticks - tablets - gaming consoles? How much will it cost me to make apps for these types of devices and how much does it cost to update them as new devices and their software versions change?

OTT Content

OTT service providers can choose to have live or stored on demand programs, media category types, and the types of distribution they will offer. Providing access to live content can be highly profitable, but requires experienced and reliable production and distribution systems. For example, a single McGreggor boxing event had gross sales of over $400 million, but there were distribution problems that resulted in very expensive subscriber refunds.

Providing access to stored video content is much less risky, but with 600+ OTT providers competing in the United States in 2019, other competitors may have similar options, so profit margins can be low. The choice of niche media categories, such as Harvard Sports TV, can have very high margins and low churn rates.

Distribution license rights for OTT content may be difficult or very expensive to obtain for certain types of movies and shows. OTT providers that have a limited amount of distribution (less than 100k subscribers) may be forced to pay much higher licensing fees or work with distributors who add their costs to the licensing.

OTT Services

OTT services can be provided on a monthly subscription video on demand (SVOD), pay per view - transaction video on demand (TVOD) or as an advertising video on demand (AVOD) service. Some OTT systems offer a mix of service options, providing basic content access on a subscription basis and small extra fees for premium content. The decision in terms of types of services to offer may be influenced by the content licensing cost model.

Some content owners provide rights to distribute their content over a period of time for a fixed fee, which is good for SVOD. This is why you may see the same movie frequently offered on a premium channel for one month.

Others may provide content on a revenue sharing basis, which is good for pay per view TVOD. This is why you tend to see pay per view premium content offers in the initial menu display pay for the OTT service app.

OTT Platforms

OTT platforms gather, organize and stream or transfer content to users. OTT service provider platform choices range from ready to use (turn-key) systems to full custom systems. The choice of ready made solutions typically comes at a much higher cost and has limited customization capabilities. Some OTT companies start out with a ready to use platform and gradually migrate to semi-custom or full custom solutions. For example, NetFlix started video storage and distribution using Amazon's AWS services. As the NetFlix network grew, the added overhead and complexity of using AWS resulted in NetFlix setting up their own content storage and distribution systems.

OTT Workflow

To get, set up, proces, and schedule content for OTT services is called workflow. OTT content workflow systems have been transitioning from dedicated workstations and local production staff to cloud based workflow management services and to contractors, freelancers and online project marketplaces. Cloud based production and workflow add flexibility and shift capital costs to operational costs.

OTT Viewing Devices

OTT viewing devices are connected media displays or adapters that can request, access and interact with content. OTT providers typically want to make their TV services available on all types of devices, including smart TVs, tablets, smartphones, gaming consoles and most other types of connected devices. Unfortunately, it tends to cost $50k or more to create an OTT app, and maintenance costs can be much higher.

OTT apps must have versions that can work with specific types of device operating systems and different features and capabilities, such as screen size, touch screens and accessories. Each time new device software and features are added, apps need to be updated or they may not work correctly.

An inexpensive way to get started is to offer a web app that looks like an OTT app, but actually works through a device's web browser. This allows the OTT service to work on most types of devices. Unfortunately, this means that some of the advanced device features available in apps cannot be used on web browser versions.

OTT Subscribers

OTT subscribers are users that can access OTT services. OTT providers define the service plans (packages, rates and territories), features and user interface, and determine how the customer can get support (customer care, self service, etc).

OTT service pricing is typically set, or strongly influenced, by the marketplace. If competitors offer similar services, it can be difficult to sell the same services at a higher price. Service territories can be set by controlling access of internet addresses (IP addresses). IP addresses are assigned by location, and it is possible to restrict IP address access by territory (geofencing).

OTT system features are often determined by the type of platform and customization projects. Essential features such as electronic program guides (EPSs) are typically available on all platforms, but may require expensive data licensing. Program and show recommendation program guides, movie restart, and other features may not be available or may not be customizable on turn key platforms. Features such as closed captioning may be required by law.

OTT customer management is a mix of service packages and support that is provided to OTT customers. Setting up and reducing customer support costs is very important to profitability. In 2019, NetFlix had 151M subscribers (Statista) and 7100+ employees (Forbes), which is a subscriber to staff ratio of 74,000 customers for each employee.

It is important to rapidly develop digital Customer Care (eCare) systems. Providing eCare self-services can improve customer satisfaction and reduce customer care support costs. Basically, helping your customers to do self-service makes them happier and reduces your costs.

OSP Business

OTT key choices include content offerings, production & distribution platforms and service packages.

Content choices include original production, repackaged media or licensed content for distribution. OTT platform options include ready to customize and use ready to customize white label platforms, open systems, and custom systems and software. Services include content bundles, device viewing options and user interfaces.

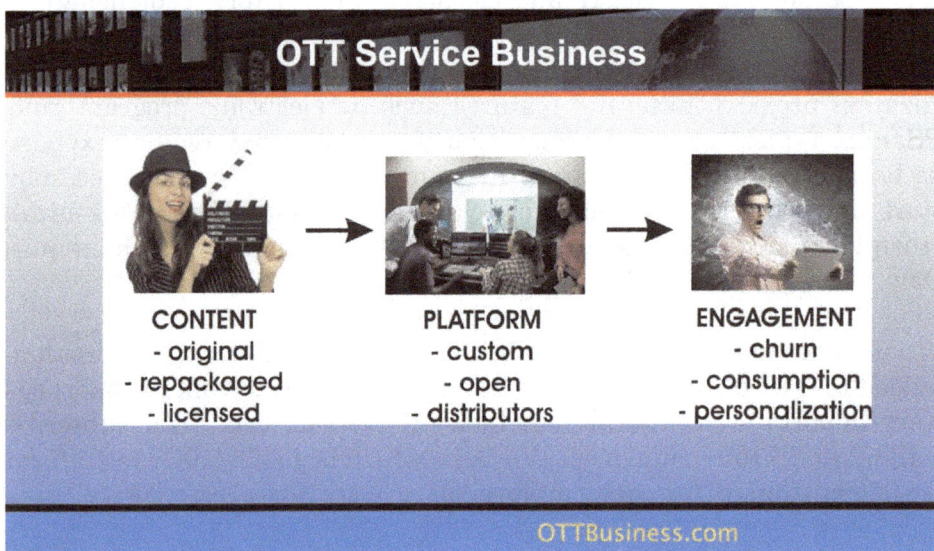

OTT Marketplace

The OTT service marketplace in 2020 had over $46B revenue, which is growing at 18%+ per year.

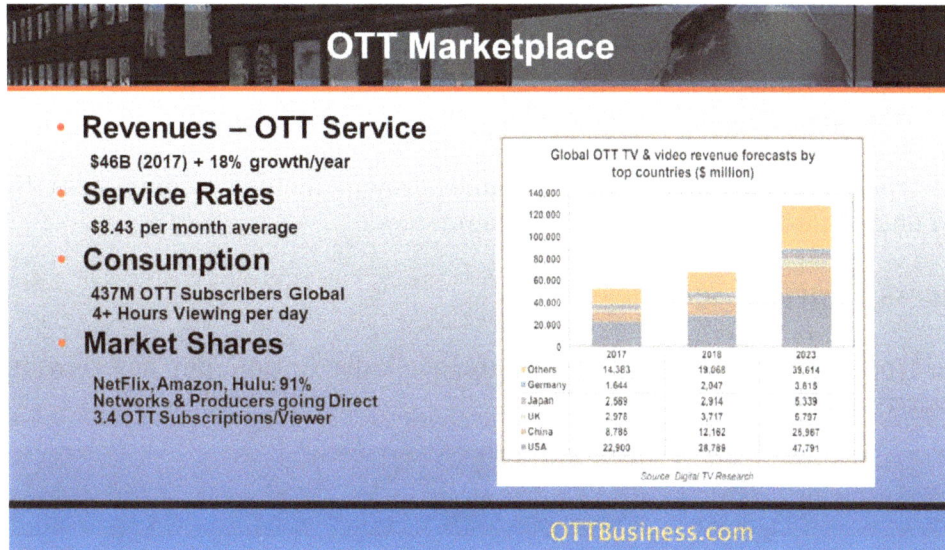

The average OTT subscription rate was $8.43 per month, with the OTT customers subscribing to 3.4 OTT services in the USA.

The OTT video ad insertion rates range from about $3 cost per thousand (cpm) for general network video ads to $28+ cpm for targeted ads. For example, if an OTT provider charges $10 cpm for ad insertion, the advertiser would pay $50 for 5,000 ad insertions.

In 2020, TV viewers watched about 4 hours of TV per day, and there were over 437M OTT subscribers worldwide. Viewer habits are changing. People are gravitating towards specialized programming and shows where there is audience engagement on and off-screen so they can interact with friends and fans. how companies are earning revenue from television services. The TV viewing time on big screens is 3x higher than mobile devices.

In 2019, the top three OTT service providers, NetFlix, Amazon Prime and Hulu, controlled 91% of subscribers in the USA. This is changing. Pay TV platforms - HBO Now, Starz, Showtime - TV networks & broadcasters (CBS) and sports networks (MLB.tv). Many networks, such as Starz and HBO, as well as content producers like Disney, are starting to provide direct OTT services. New niche OTT providers, such as Hopster TV, CuriosityStream, TubiTV and others, have achieved over 10M paying subscribers with 60% and higher annual growth rates.

OTT Services

The types of services that OTT companies provide include media access (live and on demand), advertising and tCommerce.

Access Viewing Services

The types of access viewing services include subscription, pay per view and advertising.

Subscription Video on Demand (SVOD) service requires a viewer to set up a subscription service (pre-authorization), so they can get the ability to access content for viewing. Subscription fees may be comprised of a periodic (e.g. monthly) fee and/or a per movie access fee (pay per view).

Transactional Video on Demand (TVOD) is a service that requires the viewer to pay a fee per view to access content such as movies or TV shows. TVOD service is commonly called pay per view (PPV).

Advertising Video on Demand (AVOD) is a viewing service that allows users to access content in return for viewing ads during the playout of the content. AVOD services may be provided for unregistered guests as well as registered users. Because the profile of viewers is known and ads can be better matched to viewer interests, registered AVOD services can generate much higher ad revenues for the OTT service provider.

OTT Advertising Services

OTT Advertising includes traditional types of TV video advertising and new types of digital advertising.

Traditional broadcast video ad insertion can be enhanced with targeted dynamic ad insertion (DAI) and ad personalization. This increases the value to the viewer by selecting ads that match viewer preferences,which increases the value to the advertiser and can dramatically increase the ad revenues for the OTT provider.

OTT eCommerce Services

OTT eCommerce services can include Infomercials and direct marketing. Infomercials are promotional shows or content that can be provided on a fee or revenue share basis.

Direct marketing can use the OTT show platform for direct product sales. This can include on platform video promotion or direct marketing to show subscriber lists. An example of direct marketing is how QVC or HSN shopping channels may directly promote to their viewers by email.

OTT Content

OTT content can be original content, live event, licensed content or affiliate program distribution.

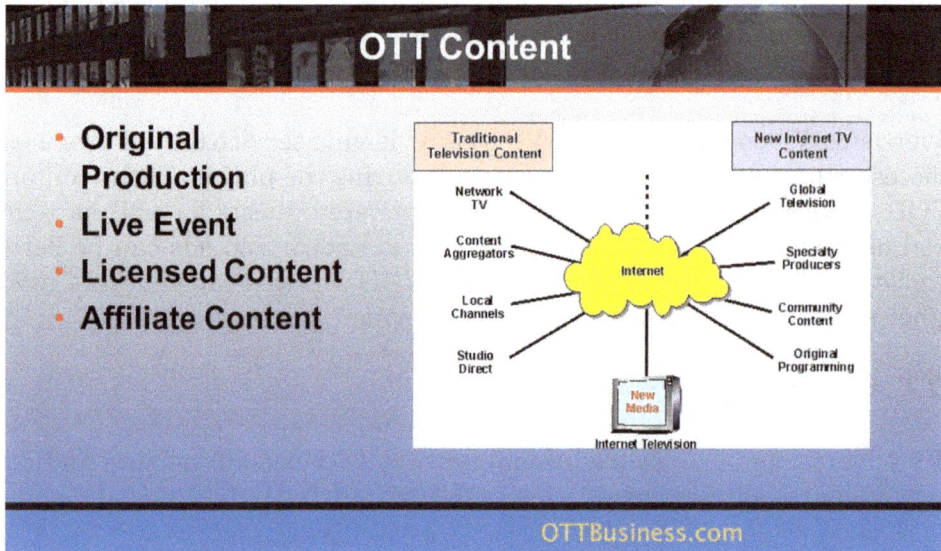

Original Production Content

Original production content involves creating new show such as sitcoms, talk shows, news and other media which may be owned by the OSP. Producing and owning content can provide long- term licensing revenue. Types of content such as the popular show series, Game of Thrones, can generate revenue for decades, while other types of content, such as news, tends to have short lifetime value.

Live Content

Live content may be from on-site production and/or connection to video feeds from live events, like concerts and sports.

An example of the high value potential for live event content is the McGreggor vs. Mayweater boxing fight in 2017. This one event generated $400 million in revenue by getting four million people to buy tickets priced at $100.

OTT Live Event Content Example:

McGreggor vs. Mayweather 2017 Fight
$400M+ Revenue
4+ Million Viewers
$100 per Ticket

OTT Licensed Content

OTT can license content rights so it can stream content such as movies, TV shows or other forms of media owned by another company.

Licensing premium content for OTT systems can be costly and difficult because content owners want to maximize the content value and protect it from being copied. Content owners may require verifiable security systems be used for OTT content delivery.

The licensing rights terms can include geographic distribution areas which can be managed by IP address location, called Geofencing. OTT systems use Geofencing to block the use of IP addresses from locations outside the authorized areas. When attempts are made to access content, from another country, for example, the connection can be refused.

OTT Affiliate Content

OTT providers may set up affiliate distribution deals that allow video content provided by another source such as TV networks, movie distributors and other media providers.

OTT Content Sources

OTT providers can get their content from content producers (studios), brand/event owners (live), distributors (movies & shows), and networks. It may be possible to get the same content (such as movies) from different sources.

Content Producers

Content producers are companies such as studios, video production companies and other organizations that create and own movies, TV shows or other media.

Brand and Event Owners

Brand and event owners organizations own the rights to event brands and activities, which can include sports team franchises or event hosts, like concerts or competitions.

Media Distributors

Media distributors are companies that obtain the rights to license and distribute content such as movies, TV shows and other media. Content producers and other sources of media use distributors so they don't need to manage communication with many licensees.

Distributors can have distributor requirements which include verifiable asset management and security systems.

Media Networks

Networks are companies that broadcast or provide media to local TV stations or other consumer media delivery systems. Networks can set up affiliate distribution agreements that require the OTT provider to stream on a revenue sharing basis or insertion of ads from the network content provider (ad holdbacks).

Public Domain Media

Public domain media are programs that have expired copyrights or have been made publicly available for free distribution.

It is important to understand that when a public domain program has been processed, a new copyright is created which requires permission from the new producer. Just because you have a public domain movie file may not mean you are allowed to broadcast or stream it.

OTT Platforms

OTT platforms are the systems and services that enable the gathering, processing and streaming or distribution of media programs and services. OTT platform options include custom, hybrid, open and white label systems.

Custom OTT Platforms

Custom OTT platforms can use any mix of computer storage & streaming servers. They also use custom software that can be set up to provide a variety of different features.

Hybrid OTT Platforms

Hybrid OTT platforms use a mix of standard equipment and customizable software, which can provide enhanced and unique services and features. An example of Hybrid OTT is the Hybrid broadcast broadband TV system.

Open OTT Platforms

Open OTT platforms are free to use software that can be customized to provide new features and services. Because they open software, new features that the OTT provider creates must be added to the open source software, allowing other companies to use the new features. An example of the open OTT platform includes the Kaltura open source platform.

White Label OTT Platforms

White label platforms are systems that are already set up and can have the appearance and features changed to represent another company by showing its brand logo, for example. An example of white label platforms is muvi.com, which has OTT platforms ready to customize and use.

Hybrid TV

Hybrid TV is the combination of broadcast television and broadband internet connection where each service connection can be used to direct, enhance and provide interactive services.

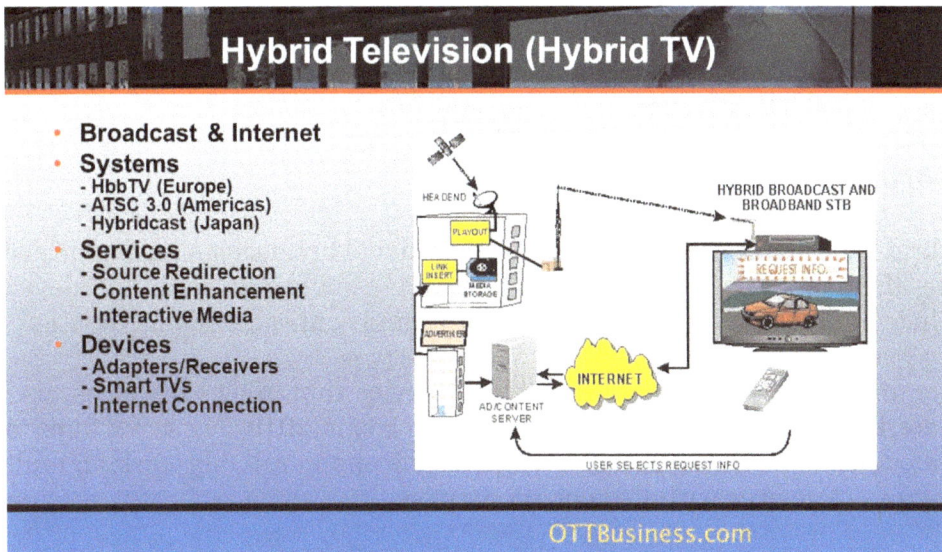

Broadcast and Broadband TV

Broadcast and broadband TV systems can modify broadcast signals to include hyperlinks which enable receiving devices to update graphics and redirect sources, such as new streaming channels or interactive services.

There are several versions of hybrid TV. Hybrid broadcast and broadband TV (HbbTV) began in Europe in 2009. HbbTV has spread to many parts of the world with 500M+ viewing devices in 2020. The Hybridcast system started in Japan in 2014 and had over 12M devices in 2020. Advanced Television Systems Committee (ATSC) version 3.0 was developed in North America, with first commercial deployment in South Korea in 2017.

Hybrid TV Standards:

HbbTV - Europe - 2009 - 500M+ Devices
Hybridcast - Japan - 2014 - 12M+ Devices
ATSC 3.0 - North America - 2017

Hybrid TV Services

Some HbbTV services include video on demand, catch-up services, electronic program guides, interactive advertising, content personalization, voting, games, social TV, TV apps and other multimedia applications.

Hybrid TV Devices

Hybrid TV Devices are able to receive traditional broadcast signals and combine them with new internet connection capabilities.Types of devices include adapters and receivers, smart TVs and connected media devices.

Operator App (OpApp) is a new HbbTV service that allows broadcasters to create TV apps that do not require a set top box in 2019. These TV apps are created using standard HTML5, allowing the app to reliably work on multiple types of devices with little or no changes.

OTT User Experience - UX

The OTT user experience (UX) includes user interface (UI), media services & features and customer support. The UX is a key value that most OSPs can provide to engage and keep subscribers.

Many of the churn management companies focus on measuring the user experience and their media consumption. Content consumption can be optimized and most of the customer churn can be predicated and managed based on use patterns.

User Interface (UI)

The user interface is the portion of equipment or operating system that allows it to interact with the user. The UI is a collection of presentation media and processes that define how an end user can interact with an application or service. The types of devices, software and systems can determine or limit the interface options with the user.

Content Value

Content value is related to audience interests, media scarcity and engagement rates. Subscribers are often willing to tolerate poor quality and features if they are getting high value content.

A good example of high content value is the success major league baseball (MLB.com) had online in the early 2000s. Subscribers were willing to pay $3 to view each game that was streamed at less than 500 kbps. In their first year, MLB.com earned over $50M from streaming services.

Content Value Example:

Early 2000s
MLB Baseball Games
Low Quality - Less Than 500kbps
$3 per Game
$50M+ Extra Revenue

OTT Content Navigation

OTT providers offer channel services and program content packages, viewing features, social sharing and navigation features. Viewing features include media player controls such as pause, skip, add to watch list and other viewing options. Social sharing allows and can encourage viewers to comment and share program media on social networks. Navigation can include interactive program guides and recommendation engines.

Customer Support

Customer support helps users obtain, set up and get answers to questions. Customer help can be provided by a mix of live help and automated help services. A measure of customer support effectiveness can be seen by subscriber to staff ratios, which range from several hundred subscribers for small and growing OSPs to over 200,000 subscribers per staff for the company Curiositystream (documentaries).

Subscriber to Staff Ratios:

Costly - 100 Subs per Staff - Example: A New Small OTT Provider
Efficient - 200k+ Subs per Staff - Example: Curiositystream

OTT Churn Management

Churn management involves predicting and managing customer service cancelling. Churn can be predicted by monitoring consumption activities such as viewing time. Churn management measures the subscriber value to determine which customers are good to keep, and which should be let go. Churn can be reduced by creating retention programs that can be provided before and during churn activities.

OTT Security

OTT systems need security processes to protect their services, maintain privacy for subscribers and meet the requirements of content providers such as studios, networks and other companies.

OTT Security Requirements

There are multiple OTT security requirements. Vendors, such as studios, and media distributors usually require verifiable content protection systems before providing content for distribution on your OTT system. Legal security requirements include subscriber privacy and content type restrictions. Business requirements include distribution and royalty reporting verification, as well as liability insurance in case of losses that your system and business can incur.

OTT Security Types

OTT security types include digital rights management (DRM), encryption and watermarking.

Digital rights management is used to assign and track content use rights, access viewing control to provide copy protection.

Encryption is the use of keys to process media so it can't be viewed or used by unauthorized recipients.

Watermarking is a process of adding or modifying information media programs so it contains content source identification. Watermarking can be for individual streams, allowing for the rapid detection and shutdown of unauthorized redistribution.

DRM systems can be provided as software that is installed on the OTT system or as cloud software as a service (SaaS). OTT systems may use multiple types of DRM services, commands and processes (protocols), to be able to work with different types of devices.

DRM services can be provided on a per use license basis, where each encryption and decryption of content for a user has a license. Licenses are provided (sold) in bundles ranging from about $10,000 to millions of dollars.

OTT Viewing Devices

OTT viewing devices convert streaming or downloaded (on demand) media into formats that can be viewed on TVs, computers and other types of media devices.

Smart TVs

Smart TVs are televisions that are connected to the internet and can run TV apps. The apps may be private (provided by the Smart TV manufacturer), or available from public App marketplaces.

Digital Media Adapters (DMAs)

Digital Media Adapters (DMAs) are media players, media sticks, connected media players, set top boxes (STBs) that include TV app capability,

Smartphones

Smartphones and tablets are media players with many display sizes and navigation options. The key types of Smartphone operating systems are Android and iOS. Smartphones have data connection and media processing capabilities and limitations.

Desktop Computers

Desktops are computers that access OTT services using web browsers with media plugins. Desktops are a relatively small percentage of streaming TV viewing devices.

Gaming Devices

Gaming devices are proprietary media player platforms. Gaming app developers need to be licensed by gaming platforms to produce TV apps, which are important for younger audiences when utilized on gaming devices.

Key OTT Companies

Key OTT company types include streaming TV service providers (NetFlix, etc), platform providers (Brightcove, etc), access device manufacturers (Roku, etc) and security systems.

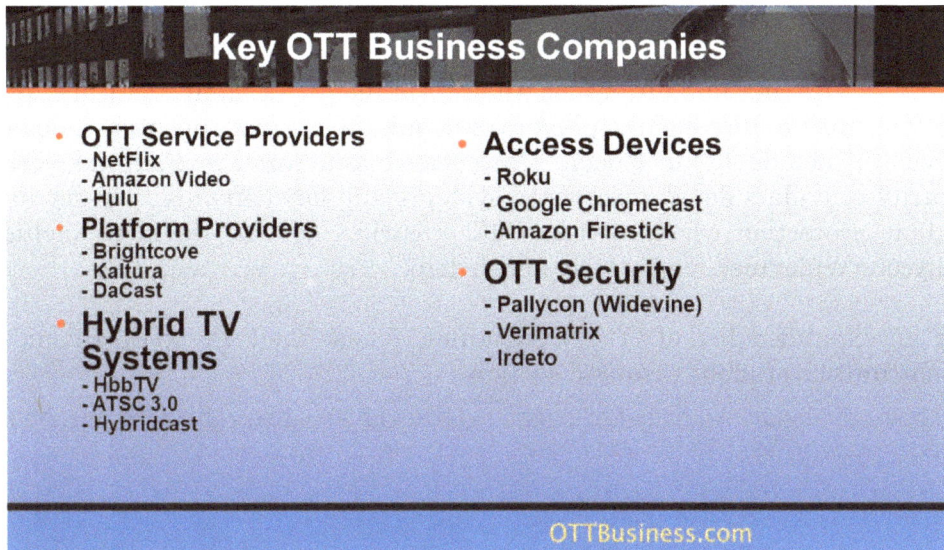

Key OTT Business Companies

- **OTT Service Providers**
  - NetFlix
  - Amazon Video
  - Hulu
- **Platform Providers**
  - Brightcove
  - Kaltura
  - DaCast
- **Hybrid TV Systems**
  - HbbTV
  - ATSC 3.0
  - Hybridcast

- **Access Devices**
  - Roku
  - Google Chromecast
  - Amazon Firestick
- **OTT Security**
  - Pallycon (Widevine)
  - Verimatrix
  - Irdeto

OTTBusiness.com

OTT Service Providers (OSPs)

OTT service providers (OSPs) supply streaming programs directly to consumers. There are thousands of OSPs, and there is a growing trend of successful niche content providers. Top providers include NetFlix, Amazon Video and Hulu.

OTT Platform Providers

OTT platform providers set up systems and software that store and stream video content for OSPs. Top OTT Platform providers include Brightcove, Kaltura and DaCast.

Hybrid TV Systems

Hybrid TV systems combine broadcast television and broadband TV (OTT) services. According to industry standards, the top Hybrid TV systems include HbbTV, ATSC 3.0 and Hybridcast.

Access Device Manufacturers

Access device manufacturers or their original equipment manufacturer (OEM) who produce media players and adapters. The top access device providers include Roku, Google Chromecast and Amazon Firestick. OTT security providers develop software systems and key software, privacy and content protection services. Top OTT security system providers include Pallycon (Widevine), Verimatrix and Irdeto.

For an expanded list of OTT Companies, please visit the OTT Business Opportunities student resource section.

Chapter 3

OTT Advertising

OTT and digital advertising is the insertion of promotional videos, graphics and other media into programs, websites and other content the OSP publishes.

OTT service providers can earn more money from advertising than traditional TV service providers by micro ad targeting, dynamic ad insertion, ad personalization and by providing cross-platform advertising campaigns.

OTT Advertising

- OTT Advertising Options
- Ad Insertion & Targeting
- Digital Advertising
- Multi-Platforms Advertising
- Ad Campaign Management

OTTBusiness.com

What do OTT companies want to know about advertising?

OTT Advertising Options - What types of ads can I insert? Where can ads be inserted? How many ads will viewers tolerate before unsubscribing?

Ad Insertion & Targeting - What systems and services do I need for ad insertion? How can ad systems target viewers by their profiles and interests? How much more revenue value can targeted ads provide?

Digital Advertising - What types of graphic, text and other digital media ads can I insert? Where can I insert digital ads? How can I automatically get digital ads to insert?

Multi-Platform Advertising - How can I offer my advertisers ad campaigns on other systems, such as Youtube, Pinterest and other platforms? How do I link the advertising systems together? How should I manage advertising inventory and billing on multiple networks?

Ad Campaign Management - How do we sell ad services and track the ads that were delivered? How do we bill and provide ad insertion reports for our advertisers?

OTT Advertising Business

Key OTT advertising options include cross channel advertising, dynamic ad insertion and personalized advertising.

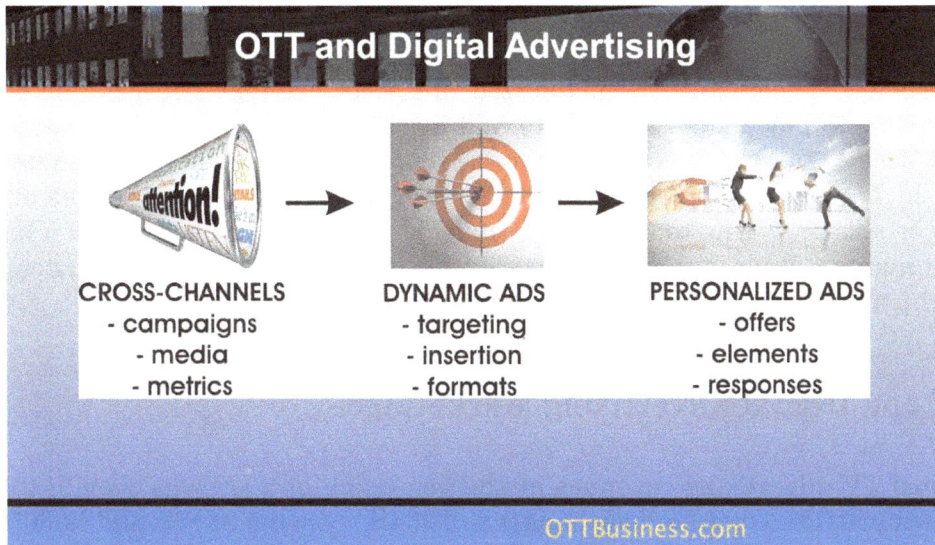

Cross-Channel Advertising

Cross-channel advertising is the running of ad campaigns that are published on other media channels such as Youtube, Google Search and other OTT provider platforms. When OTT providers have an advertiser, it is relatively easy to sell them related advertising on other platforms. In many cases, the same or slightly modified ad media can be used for ad campaigns that are run on other platforms.

Dynamic Ad Insertion (DAI)

Dynamic Ad Insertion (DAI) is the selection, delivery and presentation of ads to reach viewers with profiles that include interests and activities that match the ad campaign objectives. Targeted DAI can dramatically increase the success of ad campaigns, resulting in much higher ad revenues for the OTT provider.

DAI works by sending an ad insertion opportunity to an ad network. The insertion request includes viewer targeting information. The advertising network may provide back multiple ad options along with the ad insert rates the advertisers are willing to pay. The OTT advertising system then selects and inserts an ad into a program or other media channel.

Personalized Advertising

Personalized ads can use customized offers, insert text and graphic elements, and may enable immediate and interactive responses. Personalized ads, such as inserting a viewers favorite sports team logos into ads, can have the highest conversion success rates. However, digital media agencies may not be able or willing to set up personalized ad campaigns because of a limited number of places that offer the options, and the extra time it takes to set up and manage personalized ad campaigns.

TV and Digital Advertising Marketplace

Global TV advertising revenues at the beginning of 2020 was over $140B, growing at 1%. Digital advertising was $284B, growing at 8%, and is 60%+ of total ad spending.

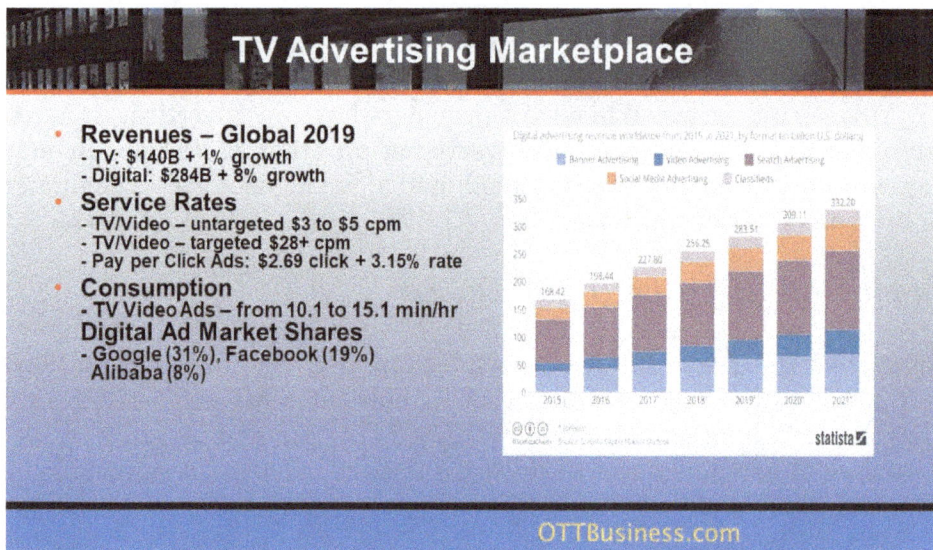

TV advertising budgets are shifting from broadcast to internet marketing, primarily because of the ability to target, measure and control ad campaigns. Ad sales are moving from direct sales agents to ad networks. OTT advertising platforms can offer better targeting, measurement and campaign control options than other types of networks.

OTT Advertising Rates

OTT advertising has multiple times the amount of revenue earning potential opportunities than traditional broadcast advertising does.

Advertising rates for untargeted video ad insertion is approximately $3-5 cpm, and targeted video ad insertion rates can be over $28 cpm.

Digital advertising can be offered at a cost per thousand cpm rate like video ads, a pay per click rate ppc, a cost per action rate cpa and other methods.

Digital Advertising Options:
- Insertion - pay when inserted
- Pay per Click (PPC)
- Cost per Action (CPA)

Google's average cost per click in 2018 was $2.69, and in 2019, Google had an ad click through rate (CTR) of 3.17%. This provides an effective cost per thousand cpm (ecpm) of over $116.

According to Adstage research, the average cost per click on the Google PPC system in 2018 was $2.76. Because only a small number of people who see ads actually click on them (about 4%), Adstage determined that the cost per impression was a little over $116 per thousand. $116 cpm is more than 10x higher than the traditional local television advertising rates of $5 - $10 cpm. The reason for this value difference is the ability to target ads to better qualified audiences. [https://blog.adstage.io/google-adwords-benchmarks-q1-2018].

Google Pay per Click (2018):
Avg Cost per Click: $2.69
Avg Click Rate: 3.17%
Effective Cost per Thousand (ECPM): $116+
Source: Adstage Research

OTT providers can offer dynamic ad insertion (DAI) and personalized ads that enable OTT service providers to earn $100+ cpm on some ad campaigns.

The amount of consumption of ads is determined by the willingness of viewers to watch ads. The amount of time that can be allocated to video advertising is called video ad load, which for cable TV networks ranges from 10.1 ad min/hr (NBCU) to 15.1 ad min/hr (Viacom) and is increasing.

Digital Advertising Market Shares

Digital advertising market shares are Google 31%, Facebook 19% and Alibaba 9%.

A key reason the TV advertising marketplace has been shifting from broadcast to internet marketing is because of the ability to target, measure and control campaigns.

OTT Ad Types

OTT Ad Types include video, graphic, audio, text and interactive.

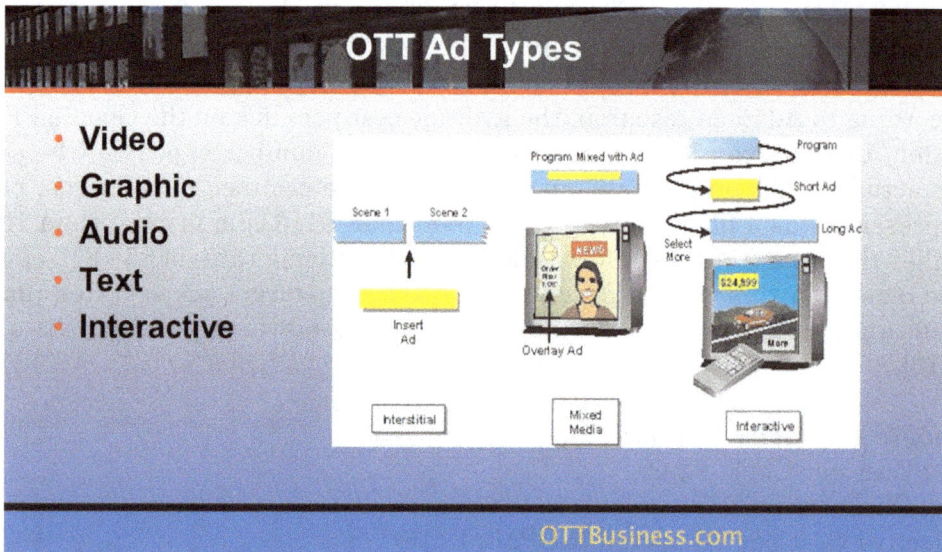

Video Ads

Video ads are typically 15, 30, 45 and 60 seconds long. The may be offered as pre-roll, inserted somewhere in the program content (interstitial) or interactive. Video ads need to have multiple aspect ratio sizes and resolution formats for distribution to different types of devices.

Pre-roll ads may include a skip option. Some companies are producing five or six second pre-roll ads to stop people from clicking on the Skip Ad Timer overlay.

Graphic Ads

Graphic ads (also called display ads) are images, logo bugs or other digital image types that may be inserted or overlaid as product placement ads.

Audio Ads

Audio ads may be inserted into audio programs such as streaming radio, talk shows and music and sports channels.
Text Ads

Text ads may be inserted into websites, apps or in search results. Text ads typically include links that viewers or visitors can click on.

Interactive Ads

Interactive ads allow users to immediately respond to ads, typically using an internet connected backchannel. Interactive responses can have extremely high value for the advertiser and OTT service provider.

Interactive ads can change or redirect viewers to other content. The immediate response option of interactive ads can provide dramatic increases in viewer conversion rates - buying inventory is very high value for advertisers. This means that OTT providers that offer interactive ads can earn significantly higher ad rates over other types of advertising.

Ad Insertion Restrictions

Certain types of ads may not be allowed, such as those containing adult content and political messages. Content owners of streaming shows or on-demand content may have requirements that program media may not be altered with graphic inserts or overlays. This helps to ensure that advertisers are not associated with content brands in the media.

OTT Ad Services

OTT advertising services can include video and digital ads on OTT platforms through partners and ad networks.

Video Advertising

Video advertising services can include direct ad sales (for specific shows and time slots), ad agency sales or may be offered to a programmatic (rules based) ad insertion system. OTT direct ad sales tend to earn higher advertising rates.

Digital Advertising

Digital advertising is the insertion of video clips, graphics, text or other digital media onto the OTT website platform, partner platforms and ad publishing networks.

Ad Networks

Ad Networks are advertising platforms that connect advertisers with publishers. Ad networks use rules based ad campaigns to gather and insert ads when matching insertion opportunities become available.

Digital Advertising Platforms

Digital advertising platforms are systems and gateways that can connect to multiple publishing networks. Digital advertising platforms enable OTT providers to sell multiple advertising services and manage campaigns on multiple platforms and systems.

Ad Campaigns

Ad campaigns include ad selection, insertion rules and budgets used by advertisers to control which ads are shown to certain types of people at allowable costs.

Advertising Sales

Advertising sales can be done by staff, sales agents, agencies, or can be automatically sold on ad network platforms.

OTT providers tend not to have skilled ad sales staff, and the amount of ad sales inventory and ad revenue potential may not be enough to hire qualified ad salespeople. The way small broadcasters have solved this challenge is by using ad sales services such as Viamedia, which uses qualified sales people to sell advertising for multiple small to medium size broadcasters.

Ad Networks

Ad networks link advertisers and publishers (OTT systems are publishers) by using ad campaign management services.

Ad networks gather ads and campaign rules from advertisers, and use these rules to provide and track the insertion of ads on publisher platforms and channels.

Advertisers

Advertisers are companies that allocate their promotion budgets over multiple advertising types and platforms. Advertisers want their ads to reach the right audiences and not be shown or associated with certain topics such as hate, violence, adult content and other bad media.

Media Agencies

Media Agencies are companies that help advertisers to create promotional messages, ad content and manage advertising campaigns. Media agencies typically act on behalf of their client companies to set up ad campaigns, buy media and optimize the marketing campaigns.

Media agencies typically earn 10% to 15% commission on advertising purchased from broadcasters and publishers. The media agency does not send a bill to the advertiser. They are paid for by the hidden fee, making it appear like there is no cost for using an ad agency.

Ad Campaign Rules

Ad campaigns include sets of rules and budgets for requesting ad insertion or publishing. There are typically multiple ad campaigns for advertising projects.

Publishers (OTT Providers)

Publishers are media companies that provide content, such as OTT streaming and media portals, to viewers and users. Publishers can be OSPs, TV broadcasters, shared media networks or any other platform or service that enables audiences to receive or get access to media.

Ad Monitoring and Measurement

Monitoring and Measurement companies gather and verify successful ad insertions. In some cases, they can measure conversion rates that link ad views with purchases.

Measurement companies can be big and costly, such as Nielsen ratings, or free, like Google Insights and Google Alerts.

Ad campaign measurements that can be monitored in real time and rapidly changed over time are highly valuable. Because OTT systems have direct connections with each subscriber, ad offers, insertions and engagements can be monitored, measured, analyzed and continually optimized. This is a major advantage to advertising that is delivered on broadcast TV systems.

Ad Targeting

Ad targeting is the delivering of promotional messages to specific types of people using a mix of viewer profile criteria.

Viewer Profiling

Viewer profiling is the targeting for desired characteristics of customer or audience. This can include demographics, behavioral and other categories.

OTT providers can link their advertising systems to third party data providers such as Equifax, Amazon and other companies that have a history of the types of products and services people purchase and use. Using these types of third party data management platforms (DMP systems), it is possible to target profile data without releasing private profile information to the advertiser or OSP.

Ad Selection

Ad Selection includes identifying and requesting candidate ads and their ad insertion rate offers. A campaign can have multiple ads and promotional offers.

Ad Insertion

Ad insertion is the process of selecting and splicing, or merging ads, into media streams.

Ad Measurement

Ad measurement is the gathering of ad insertion, consumption (e.g. amount of viewing time) and engagement activities, such as clicks or contact submissions.

Ad Attribution

Advertisers want to understand which ads, platforms and combinations influence purchasing or desired results. This is called ad attribution.

Dynamic Ad Insertion (DAI)

Dynamic ad insertion (DAI) selects and splices ads into media streams or programs when advertising placement opportunities become available.

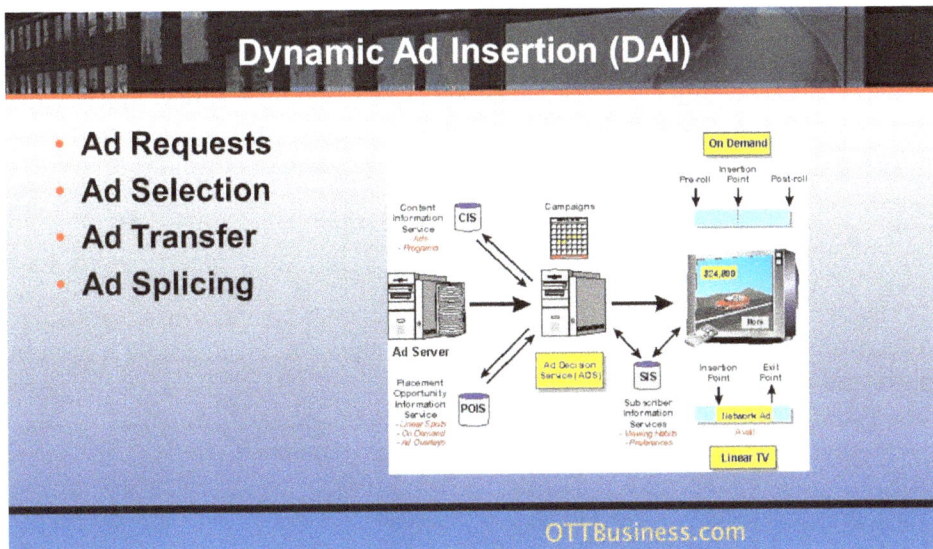

DAI ad selection and insertion criteria can include previous viewer engagement activities combined with additional information from other data management platforms (DMPs).

To enable DAI, protocols (processes and commands) have been created that allow requesting, delivering and tracking the insertion of ads. These include video ad serving template (VAST) and other supporting protocols.

Ad Requests

Broadcasters and publishers send requests for ads when ad insertion opportunities become available. In general, it is relatively easy to do DAI for a small number of users. It is much more challenging to do DAI for a large number of viewers, thousands to hundreds of thousands of which may all be getting dynamically inserted ads at the same time.

Dynamic ads inserted inside the OSP platform are called server side ad insertion (SSAI), which is undetectable by the viewer. Ad insertion done inside the viewing device is called client side ad insertion (CSAI), which may allow for ad skipping.

It is possible to pre-load ads in a viewing device, such as a media adapter, or set top box (STB) to eliminate the ad delivery setup time.

Ad insertion opportunities can be managed by supply side platforms known as SSPs, which provide available commercial insertion times and lengths. Ad insertion offers can be managed by demand side platforms called DSPs. Ad insertion rules and offered rates are provided when requested.

Ad Transfer

Ad transfer is the moving of the ad from ad servers to streaming platforms. It takes time to request, transfer and process the ads. This can be a big challenge for ad insertion into live media streams.

Ad Splicing

Ad splicing is the insertion of ads into a video stream or replacing existing program content. For digital video, this can be a complicated process because compressed video contains keyframes that occur every 1-2 seconds. Ad insertion requires modifying the media to create new keyframes (media stream processing), where the ads can be inserted.

Ad Retargeting

Ad retargeting is the identification of a person who is watching certain types of ads and sending the same ad or related ads to the same person.

Ad Retargeting

- **Ad Remarketing**
- **Re-Targeting Value**
  - recognition
  - credibility
  - conversions
- **Recipient Identification**
- **Ad Redistribution**

RETARGETING

EMAIL
SOCIAL
SITE
SEO/SEM
SEARCH
CONTEXTUAL

OTTBusiness.com

You may have noticed ad retargeting as you browse the internet. Ads seem to follow you as you go to different websites. If you visit a website that sells washing machines, you may be tagged as a visitor having interest in washing machines. Even when you visit a car dealership site, for example, you will continue to see ads for washing machines. The increased frequency of ads can dramatically increase the viewer's memory of the ad and interest in responding to it.

Ad Remarketing Campaigns

OTT service providers can set up ad remarketing campaigns that allow advertisers to define which ads should be redistributed and how many times they can be displayed.

Ad retargeting provides higher value for the advertiser and more ad revenue for the OTT service provider. Viewers develop awareness for ads they see (recognize) multiple times. Seeing repeated ads (ad frequency), tends to have higher conversion rates.

Ad Recipient Identification

Ad recipient viewer identification can be done by tracking a first time ad view and saving (pushing) an ad ID cookie in the viewing device memory. This allows the advertising system to determine that a specific ad has already been viewed by the person.

Ad Redistribution

Ad redistribution is the sending of the same or related ads to a viewer who has already seen an ad. The next time this viewer's browser or device has an ad insertion opportunity, the advertising system reviews the viewer history or an ad cookie to determine the types of ads to show. If the ad history is part of a remarketing campaign, the ad system can select and insert the same or a similar type of ad.

Ad Personalization

Ad Personalization is the modification of advertising message elements, including text, images and audio to match the interests or behaviors of the ad recipient.

The ad personalization modifications can be simple, such as selecting previously modified ad variants to the dynamic creation of new ads using visual element changes. Examples of this include inserting the logo of a viewer's favorite team or putting the name of a viewer's friend in the contents of an ad.

Ad Elements

Personalized ad message elements include the offer, icons, text, graphics and audio and video clips.

Offers such as the item, price or quantity can be selected or set up to match specific customer types and interests.

Recognizable icons, text, and graphics that identify sports teams, schools and brands can be inserted to build trust and credibility. Videos can be selected or dynamically created to match interests and viewer engagement preferences.

Personalized Ad Data

Personalization data is the information that can be used to modify the offer, elements and other ad media. Personalization data can come from the user profile, OTT platform engagement activity or third party data providers to match advertiser campaign requirements.

Dynamic Video Ads

Dynamic video ads are sequences of video clips that are immediately constructed using viewer data.

Dynamic video ads can be created by combining multiple short video segments (one to five seconds each). These segments contain metadata that allow them to be selected and structured in sequence to match viewer interests. They can be immediately combined (rendered) and sent to the viewer.

An example of dynamic video personalization is the creation of political viewpoint messages. One of the companies I interviewed in 2015 provided a dynamic video production service for political fundraising that created candidate videos which matched visitor interests.

The politician made dozens of video clips about his viewpoints. Segments that matched the visitor's profile and engagement activity were combined, so the resulting video matched exactly the interests of the visitor.

Digital Advertising

Digital advertising is the delivery of promotional messages on multiple types of devices (TVs, computers, smartphones), in multiple formats (video, images, text), through multiple platforms (TV, Social Media and others).

Digital Ad Services

OTT service providers and TV broadcasters can sell digital advertising in addition to video ads. Digital ads can be inserted into the OTT and Broadcaster websites, sent to multiple types devices such as mobiles, tablets and laptops, and also published on other platforms including partner systems and ad networks.

To sell digital ads, OTT providers can update their video ad sales programs to include sales agents, advertising agencies and advertising networks. OTT sales agents can earn additional sales commission by also selling digital ads.

Integrated Advertising

Integrated ad campaigns have the ability to set up campaigns that are coordinated over multiple types of devices, channels and systems.

Example of Integrated Advertising:

Each time I went to buy TV and radio advertising for my trade shows, conferences and schools, representatives from iHeartRadio, Capital Broadcasting, Time Warner (now Spectrum) and Viamedia (advertising agency), quickly started to discuss digital advertising options with me. They offered to add or expand my advertising campaigns on partner platforms, including Youtube, Google Adwords and other networks, explaining that integrated ad campaigns can be more effective and cost less to manage.

Key Advertising Companies

The key types of companies that set up and run advertising services for OTT providers include ad networks, server side platforms, data management platforms, demand side platforms, ad measurement services and dynamic video services.

Key OTT Advertising Companies

- **Ad Networks**
  - Decentrix
  - Google DFP
  - Media.net
- **Supply Side Platforms (SSPs)**
  - Google Ad Manager
  - Open X SSP
  - Pubmatic
- **Data Mgmt Platforms (DMPs)**
  - Lotame
  - Oracle DMP
  - OnAudience.com
- **Demand Side Platforms (DSPs)**
  - TubeMogul
  - Google Doubleclick
  - Rocket Fuel
- **Measurement Services**
  - Nielsen
  - Moat Analytics
  - Conviva
- **Dynamic Video Services**
  - Iris.tv
  - Udinii
  - XandR

OTTBusiness.com

Video Ad Networks

Video ad networks are marketplaces that link advertisers to publishers and allow for the automatic requesting, selection and insertion of ads. Top ad networks include Decentrix, Google DFP, Media.net, amongst others.

Ad Supply Side Platforms (SSPs)

Ad Supply Side Platforms (SSP) are systems and services that manage and request ads. Key supply side platforms include Google Ad Manager, Open X SSP and Pubmatic.

Data Management Platforms (DMPs)

Data Management Platforms (DMPs) gather, organize and use customer data for ad targeting and delivery. Top DMP companies include Lotame, Oracle DMP and OnAudience.com.

Demand Side Platforms (DSPs)

Demand Side Platforms (DSPs) are systems and services that provide ads, video, and other media formats, to publishers. Key DSP companies include TubeMogul, Google Doubleclick and Rocket Fuel.

Advertising Measurement Services

Advertising measurement service providers are companies and platforms that measure ad viewing and engagement. Top ad measurement services include Nielsen, Moat Analytics and Conviva.

Dynamic Video Services

Dynamic video service provider companies provide advanced video analytics and real time video rendering services. Top dynamic video service providers include iris.tv, XandR and Innovid.

For an expanded list of OTT and Digital Advertising Companies, visit the OTT Business Opportunities student resource area.

Chapter 4

TV and Mobile Apps

TV apps and mobile apps are software programs that run inside media viewing devices such as Smart TVs, digital streaming adapters, smartphones and media players that can be connected to the internet. Key ways to use TV apps to make revenue include using apps to provide video streaming services, in app advertising, in app purchases and the sale of the TV app.

What OTT companies want to know about TV apps:

Key App Benefits - Why create TV and mobile apps? Can I run a streaming TV business without apps? What kind of new services can I offer using TV and mobile apps?

App Development Options - How complicated is it to create an app? How long does it take to create an app? What skills are required to create TV apps? How do I find quality people or companies to create TV apps?

App Costs - How much does it cost to develop mobile and TV apps? What are typical app distribution costs? What are the costs for fixing and updating apps?

App Distribution - What are new ways to download apps into Smart TVs, media adapters, mobiles, gaming consoles? Are there app distribution platforms and how do I use them? What types of terms and deals exist for preloading apps into devices or partner distribution marketplaces?

App Advertising - How do I insert ads into mobile apps? Where do TV in-app ads come from? How much money can in-app ads earn? What kind of systems and services are needed to provide in-app advertising?

In-App Purchases - How do I insert in-app purchase offers? What types of products and services are offered in TV and mobile apps? Where do the in-app offers come from? How much money can in-app purchases make?

TV App Business

OTT providers earn revenue using TV apps by providing streaming services, in app advertising and in-app purchases.

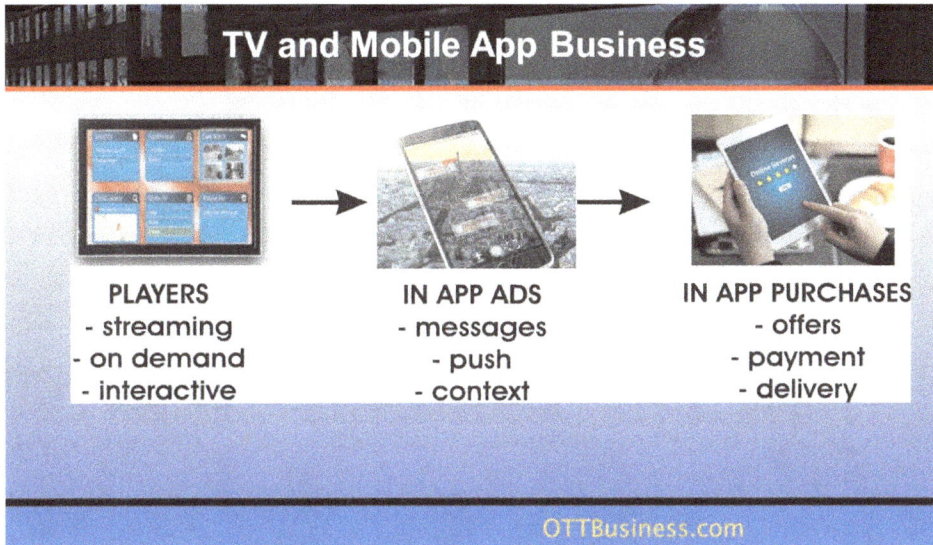

OTT app players can be used for streaming, on-demand and interactive services. TV app players need to be designed for easy install and configuration by the user in order to work with the OSP system and have an easy-to-understand user interface.

In-app ads can include offer messages and images. OTT operators can push ad media, which users typically can't block. Ad topics and offers can be selected to match the context of the content playing in the app.

In-app purchases can include direct and affiliate product offers and allow users to pay with their OTT subscriber account or other payment options. In-app purchasing systems may include integration with order processing systems, so the user can track product delivery status through the app.

TV App Marketplace

Global streaming app marketplace revenue in 2019 was $81.6B, incrementally growing at 14% per year. Mobile app market revenue was $122B and continues to grow at 18% per year.

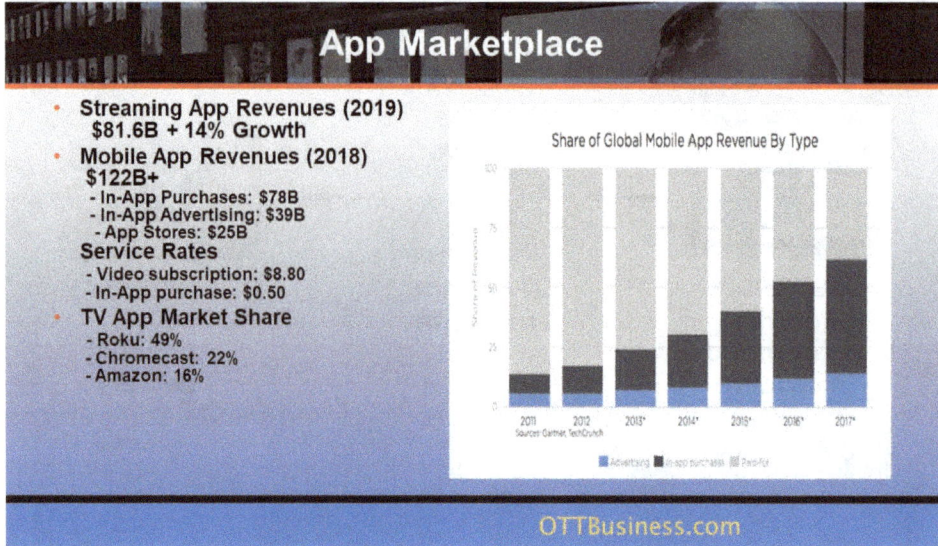

Market data from the TV app industry is hard to find, verify, and can be highly variable, depending on which research company provides it. Most app data relates to mobile apps, and little information is available about TV apps. Because connected TVs operate like smartphones, mobile app data is a good reference for predicting the TV app marketplace.

Mobile App Revenues

Out of the total $122B mobile app revenue in 2018, about $78B (~63%) was from in-app purchases, and about $39B (~31%) came from in-app advertising.

Mobile App Revenue (2018):

In-app purchases - $78B
In-app advertising - $38.9B
Source: Gartner, Techsource

App revenues can be generated through app marketplaces. $24.8B came from Google Android, and $46.6B was from Apple iOS marketplaces and directly from service providers. App marketplaces typically charge about 30% commission.

Mobile App Marketplaces:

Google Android - $24.8B
Apple iOS - $46B
App Commission - 30%

App Service Rates

The average app purchase amount (for pay to own apps) was $12.77 on iOS, $6.19 on Android, yielding an $8.80 average. IOS users typically pay more for apps than Android users. The average in-app purchase, such as for additional lives in games, is $.50 per purchase, with an average purchaser spending $9.60 per month.

Average App Purchases:

$9.60 per Month
- $12.77 on iOS
- $6.19 on Android
 -$.50 per in-app purchase
Source: AppsFlyer

TV App Market Shares

For TV app market shares, Roku is the leader, with 51% streaming market, and Fire TV with 28.5%.

TV App Market Share (2019):

Roku - 51%
Fire TV - 28.5%
Source: TDG Research

TV App Types

TV app types include proprietary, open standard and browser based web apps.

Proprietary Apps

Proprietary apps use software that only operate on specific types of systems and devices. Proprietary apps have the capability of using advanced device features and capabilities. This gives more control of the user interface to the app provider. For example, to create proprietary apps, you must be or use a licensed developer for the company that makes the devices.

Open Apps

Open apps use software that can operate on any standardized platform such as an Android. Because they are standardized and well tested, features tend to be highly reliable and interoperable. Open apps may be distributed on app marketplaces.

Web Apps (Faux Apps)

Web apps, also called faux or fake apps, are service applications on websites that appear and operate similar to TV apps. They are accessed by web browsers on viewing devices. Because they are web browsers, web apps tend to work on most types of devices. However, web based apps cannot use all of the advanced device features and users can dislike them.

Public and Private Apps

All types of apps can be public or private. Public apps can be discoverable and are directly available to everyone (e.g. in the device catalog of apps). Public apps typically require a review and approval process that can take weeks.

Private apps are manually installed and do not require reviews or approvals. They may be loaded into a device using a side loading app, which allows an application to be installed on devices directly from a web URL.

TV App Development Example:

I created a private Roku app in Brightscript programming language, which can be freely installed on any Roku player as long as I have the private app installation link for download and installation. There is no waiting period when releasing private apps.

TV App Services

TV app services include media access, user applications, in-app advertising and in-app purchases (eCommerce).

Media Streaming Apps

Media streaming app players connect a media device to the source, such as a streaming media file. To manage streaming during changing Internet connection speeds, OTT systems can use adaptive bitrate (ABR) capabilities. ABR automatically changes the quality (compression) of the video signal when it detects increases or decreases in the available Internet connection data rates

User Applications

End user apps can be used for other services and applications aside from media players. These include games, utilities and other software processes the user can directly access on his/her device, typically with or without an internet connection.

App Advertising

In-app advertising gathers and displays promotional messages inside the app. Sending messages through apps can't be blocked like email or text messages can be.

In-App Purchases (eCommerce)

In-app purchases are the processing of product or service offers and orders using mobile apps which connect users to eCommerce systems. These products may be offered directly or by affiliate partners.

TV App Development

TV app development involves selecting features and services OTT customers can get through an app on types of devices your apps can run on. Development options determine the time and cost of app creation.

App Feature Requirements

TV app features are primarily determined by the competitive marketplace by figuring out what consumers want, and what features and services competitors are offering.

App Device Types

The choice of platforms such as vendor, custom or Android, can limit or enable features and services your apps can provide. App software must have versions for different device devices including media players, smart TVs and gaming consoles. It costs money and time to adapt your apps to different types of devices.

App Development Options

App software development options include using programmers on staff, contractors or development companies to determine which languages, software development kits (SDKs), tools and software modules they should use.

App Development Costs

The cost for app development ranges from nothing or little upfront for platform provided apps that take 30% to 50% of the revenue share, to ready-to-brand white label apps that are $5k to $10k, and go up to full customizable apps, which start at $50k or more.

App Costs:

Platform App - 30% to 50% Rev Share
White Label - $5k to $10k
Custom App - $50k+

TV App Distribution

TV app distribution is the transfer and installation of TV apps into connected media devices.

In the early days of Netflix and Amazon prime, manufacturers who agreed to pre-install apps on their devices, such as DVD and media players, received new subscriber activation commissions up to $50 per new customer.

TV App Installation

To install TV and mobile apps, software is transferred into device long term memory, configured to work with the device and user account data may be added.

Pre-Installed Apps

To have apps already loaded into devices (pre-installed), software is provided to connected device manufacturers for inclusion in the device production process. This typically requires a business agreement that defines incentives the connected device manufacturer will receive for pre-installing the app on their devices.

App Marketplaces

App marketplaces are online locations where connected media devices can view, select and download applications. App marketplaces may be proprietary where the vendor controls which apps are available to the user, or they may be open, allowing users to choose any apps in the marketplace.

TV App Marketplaces:

Proprietary
Open - any apps

TV App Distributors

TV app distributors are companies that gather, manage and package TV apps for distribution to broadcasters, OTT providers, consumer device manufacturers and, ultimately, to their customers. App distributors can take a percentage of subscription and/or ad revenue for their app distribution role.

TV App Maintenance

TV app software needs to be updated when new types of devices, operating systems and new features are added.

While creating the initial app can seem expensive - $50k or more for cus-tomized media apps - a much bigger cost can be updating and testing the app software for hundreds of device types and constant updates of Android and iOS operating systems.

App Updating Requirements

App maintenance includes adding new features, such as navigation, dynam-ic ad insertion and personalization, program recommendation guides and fixing errors (bugs). Apps may need to be updated to work with new types of devices, screen sizes, user control options and accessories. Apps may need to be updated when product manufacturers release new device operating sys-tems.

App Updating Options

App updating may be automatic or be requested by a user. There may be a large number of update requests when a new version is released. Updates may be performed in groups to avoid massive negative results if there is a problem with the updating process.

App Management

Apps need to have a system that can track and manage app versions, features, compatibilities, distribution and update status.

Key TV App Companies

The key types of companies involved in TV apps include TV app developers, app marketplaces and app distributors.

Key TV App Companies

TV App Developers
- Accedo
- 24i
- Applicaster

TV App Marketplaces
- Android TV
- Apple TV
- Opera

TV App Distributors
- Zype
- Muvi
- MobiTV

OTTBusiness.com

TV App Developers

TV app developers are software development companies that specialize in creating video apps for Smart TVs, digital media adapters and other connected TV devices. Top TV app developers include Accedo, 24i and Applicaster.

TV App Marketplaces

TV app marketplaces are online portals where users can find and download TV apps. TV app marketplaces include Android TV, Apple TV, Opera and Samsung.

TV App Distributors

TV app distributors, such as Zype, Muvi and MobiTV, are companies that submit apps to multiple marketplaces.

For an expanded list of TV app companies, please visit the OTT Business Opportunities Student Resource Area.

Chapter 5

Social TV

Social TV is the merging of television broadcast and social networks to provide a viewing experience where people can find, see and/or interact with shared media.

Some of the things OTT companies want to know about Social TV include:

Audience Social Media Engagement - How to use program content and social media to attract subscribers, engage them and motivate them to stay with the OTT service?

Video Clipping Promotion - How to find, select, process and publish video clips from programs and other sources on social networks?

User Generated Content (UGC) - What are key ways to motivate people to submit valuable content? How to get permission from people to use their submitted content? Where and when to publish user generated content?

Social Ad Revenues - How to advertise on social media? What ad insertion rules to require and follow? How much social media ad revenue sharing can be earned from Facebook, Youtube and other social platforms?

Fan Management - How to build a list of fans? What are key ways to engage fans? How to earn revenue from fan activities, such as fan clubs, events and other group services?

Social TV Business

The key areas of social TV for OTT providers include content, services and fan management.

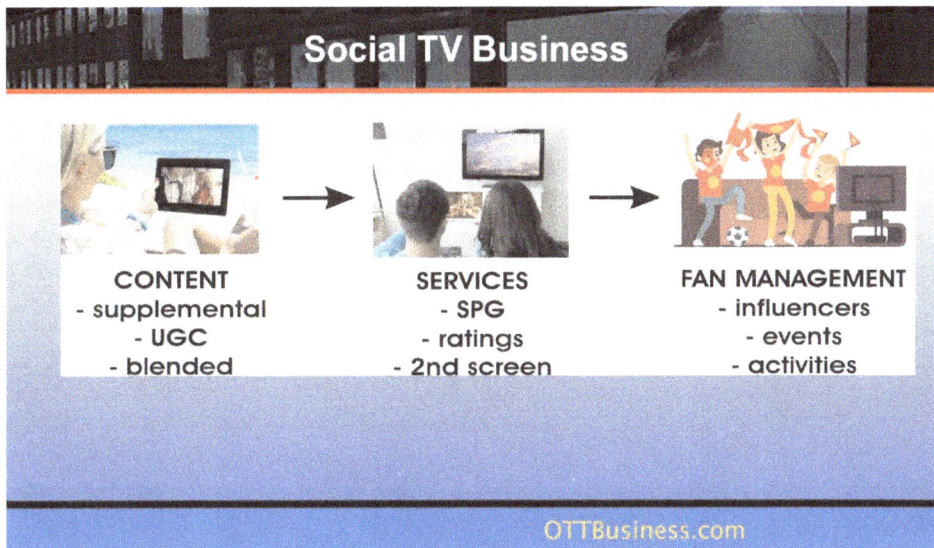

OTT providers can use social TV content and services to attract and engage viewers. Viewers may create and contribute content that can be merged for TV and promotional media.

Social TV services include social media sponsorships, new channel services and social advertising.

Fan management includes finding and working with influencers, building fan lists with profiles, working with and/or running fan events and other valuable fan activities.

Social TV Marketplace

The social media advertising marketplace in 2020 was $106B. 45%+ of viewers typically use a second screen while watching TV.

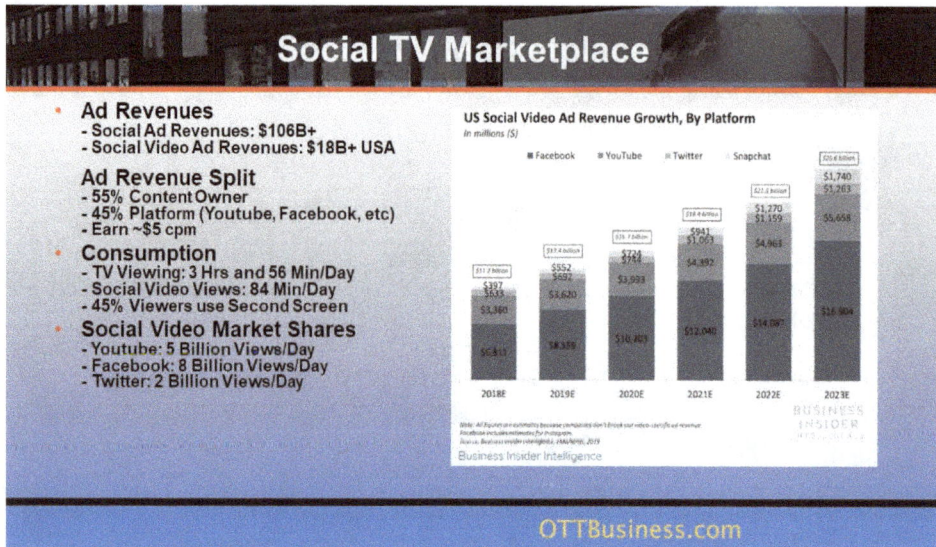

Social video advertising revenues were $106B, which comes from digital advertising, including images, messages and other media. Social video ad revenues were $18B, which comes from video pre-roll, mid-roll, and other video insertion options.

Social Media Ad Revenues

Social networks split advertising revenue for contributed videos. Typically, 55% of ad revenue goes to the content owner and 45% of it is kept by the social network platform, such as Google or Facebook. Social ad revenue typically provides the content owner with approximately $5 cpm revenue.

Video Consumption

On average, audience TV video consumption is about 4 hours per day, and social video viewing time is approximately 84 minutes per day. About 45% of TV viewers also use their 2nd screen devices during television viewing.

Social Video Market Shares

Social network video views in 2020 were about 5 billion per day for Youtube, 8 billion per day for Facebook and 2 billion per day for Twitter.

Social Video Views (2020):

Facebook - 8 Billion per Day
Youtube - 5 Billion per Day
Twitter - 2 Billion per Day

Social TV Services

Social TV services are processes that enable TV systems to gather, analyze, publish and interact with social media content and networks.

Social TV Services

- Video Clipping
- User Generated Content (UGC)
- Blended Media
- Second Screen
- TV Backchannels
- Fan Management

OTTBusiness.com

Social Video Clipping

Social video clipping involves finding interesting and motivational posts from programs and viewers, and publishing it on social channels to attract viewers and subscribers.

User Generated Content (UGC)

User generated content (UGC) is viewer contributed or shared content that can be used as marketing materials which are merged or blended into programming or combined to create new programs and media services.

Blended Media

Blended media is the combination of program content (such as movies or TV shows) with other content (such as social media or user generated content).

Second Screen

Second screen is a content distribution service that allows a TV viewer to display related information on the current program or show it on other devices.

TV Backchannels

TV backchannels are communication paths from viewers that can gather, analyze, and respond to social media and engagement activities related to a TV program.

Fan Management

Fan management includes building and classifying lists of fan contacts, monitoring and engaging in fan activities and setting up and running fan contests, events and other activities.

Social TV Video Clipping

Social TV video clipping is selecting segments of television video and processing and packaging the clips for publishing on shared media channels.

Video Clipping

Social TV clipping is used as a key promotion activity for sports and other events. It can also be used for sponsorship promotion programs.

Video Clip Discovery

Video clip discovery is the identification and selection of video segments that have value for publishing on other media channels. Discovery can come from manual review, social feedback such as comments about an event, or through video artificial intelligence (AI), which is set up to detect cheers, sports goals or other responses.

Video Clip Processing

Video clip processing is the modification and addition of media to a video segment. This includes getting permission from creators and copyright and brand owners, if necessary. Video processing usually involves reformatting the video into different sizes and resolutions so it's in a format that can be sent on to multiple media channels.

Video Clip Publishing

Video clip publishing includes selecting media channels, adding titles and descriptions, as well as formatting content for particular media channels, including a desired call to action that viewers can take. Media channels can include Youtube, discussion groups and directing content to fan group subscribers and other places.

User Generated Content (UGC)

User Generated Content (UGC) is text, images or other media that is produced by viewers, fans or other audiences.

UGC Submission

OTT operators can set up multiple ways for viewers and fans to submit content, including by forms, email, messaging or other communication methods. The submission process should include content type rules and provide permission for the OTT Operator to use the content.

Content Submission Example:

VegiPlus.tv is an MTV-type cooking show created from short cooking clips that are combined with show host introductions to produce TV show episodes. Contributors are independent producers who give permission to VegiPlus.tv to use their content. In return for submitting cooking videos, owners of the clips can receive small non-exclusive license fees, percentages of affiliate and ad revenues and may receive show talent status (profile) and other media benefits.

Content Submission Rewards:

- Small License Fee
- Percentage of Ad Sales
- Talent Status

Content Use Permissions

Content use permissions involves getting authorization from content owners, which is typically done on a content submission form. This details how their submitted content can be used. All submitted content has to be reviewed for copyright and trademarks, especially since contributors may not check or be aware of what the requirements are.

Content Review and Approval

Staff and producers view and categorize submitted clips to make sure the content is acceptable for use (e.g. no fake news or trademark violations) and determine how and where it may be used.

UGC Publishing

UGC publishing is reviewing, selecting, prioritizing and inserting/publishing UGC content into broadcast streams and/or inserting it into other media channels.

Blended Media

Blended media is the combination of program content (such as movies or TV shows) with other content (such as social media or user generated content).

Mixing Broadcast and Broadband

Blended media involves updating or enhancing broadcast or streaming media using social media content. This includes selecting program content that is a good candidate for enhancement. The next step involves discovering, gathering, processing and getting usage rights for social media that may be used in the program. Media insertion points are identified and social TV producers place social media content into the program.

Social Callouts

Social callout requests are messages inserted into media that identify, motivate and provide information on how viewers can engage with the media. This may range from the simple mention of a website to QR codes that viewers can scan that go with specific URLs.

Social Content Insertion Points

Social content insertion points are graphic areas where social media can be inserted into programs and streams. Insertion points can define the size, media formats and time intervals which are used to manage insertion opportunities.

Social TV Producer

A social TV producer is responsible for monitoring audience and fan activities, creating social media content and inserting media into shows and programs.

Second Screen

Second screen is a content distribution service that allows a TV viewer to display related information on the current program or show it on other devices.

Second Screen Content

Second screen can display the same show content (simulcast), additional media (supplemental) or related information on the current program or show, or on the screens of other devices.

Second Screen Apps

Second screen apps are software programs that link users to additional content related to a program or show. Second screen apps may include features that allow a user to view or interact with additional second screen media. This can include polls, access to exclusive fan materials or other value added content services.

Second Screen Synchronization

Second screen synchronization is the process of adjusting the relative timing of the display on a second device so it matches the exact transmission time of a primary broadcast TV or streaming channel. The process of second screen timing adjustment compensates for the typical 6 second broadcast TV delay and may need to dynamically adjust for streaming media delays.

Synchronization becomes very important in situations where the broadcast TV and second screen OTT viewing may occur in the same location. For example, watching a football game in a bar which shows the game on televisions and where some people may be watching the game on their Smartphones. If the timing isn't synchronized, the TV or the Smartphone would display events such as goals at different times - game spoiler alert!

Second Screen Monetization Options

Second screen monetization options include subscription services, which provide viewers with access to additional content such as sports statistics, interviews, background stories, along with ad revenue and in-app purchases.

TV Backchannels

TV backchannels are any type of return communication path that can be used to gather and transfer information from a viewer or service user (such as interactions, responses, user feedback, ratings, or commentary).

TV Backchannels

- Backchannel Types
- Backchannel Info
- Backchannel Monitoring
- Backchannel Communication
- Media Agents

OTTBusiness.com

TV Backchannel Types

TV backchannel information may come from OTT channel in-program options, selections or forms, interaction with Social TV platform portals (show websites, etc), and social network pages, groups or other indirect ways to communicate.

TV Backchannel Information

TV backchannel information may be directly gathered (such as links selected in a TV show) or indirectly collected (such as monitoring social media channels).

Backchannel Monitoring

Backchannel monitoring can range from setting up fan submission alerts to monitoring media channels or postings on the internet for keywords, phrases or activities. Comments and media posts may be analyzed to determine the sentiments of the media (good, bad, etc). Direct feedback submissions may trigger the sending of alerts to producers, allowing for rapid review and use of social content and contributed media.

Social Media Agents

Social media agents are people that are authorized to communicate on OTT owned or managed media channels. Media agents are typically given objectives, rules, channel list and access to content/resources to communicate. Media agents may be paid per hour, per project or per post.

Fan Management

Fan management includes building and classifying lists of fan contacts, monitoring and engaging in fan activities, setting up and running fan contests, events and other activities.

Fan Lists

Because the costs of producing and distributing shows on the Internet is now low, there are many new shows and programs being released. Building, categorizing and updating fan lists is important for attracting and engaging with show fans.

Fan Management Platforms

Fan management platforms are systems and services that gather, categorize and organize fan contacts and their profiles, manage multi-channel media communication and organize fan engagement events and sponsorship programs.

Fan Communication

Fan communication involves managing messages and campaigns such as announcements, polls and contests. It is important for fan communication to have guidelines that identify and manage unauthorized content and media that may need to be processed by different people.

Fan Content Access

Providing fans with accessibility to additional show related content, such as celebrity interviews, back stories and access to the show before it's broadcast, is a key way to build fan lists and provide and maintain high fan engagement.

Key Social TV Companies

Key Social TV company types include social TV Platforms, Social TV ratings, fan management and TV Video clipping services.

Social TV Platforms

Social TV platforms gather, organize and publish social media related to OTT channels and services. Key Social TV platform companies include Never.no, Flowics, Vizrt, Actus Digital and Make TV.

Social TV Ratings

Social TV ratings companies review, analyze and measure TV viewer media feedback engagement on social media. Key social TV ratings companies include Nielsen, Kantar and Shareable.

Fan Management

Fan management companies and platforms identify, store, categorize, manage, send messages and track fan activities. Key fan management companies are Vivoom.co, Monterosa (Lvis) and Fantrust.

TV Video Clipping

TV video clipping companies and services find, select, describe, package and publish TV video clips. Key TV video clipping companies include Graybo, Blackbird and NewTek.

For an expanded list of Social TV Companies, visit the OTT Business Opportunities Student Resource Area.

Chapter 6

Television Commerce (tCommerce)

Television commerce (tCommerce) is the providing of electronic commerce services through television systems.

What OTT companies want to know about doing tCommerce:

How to do eCommerce on OTT Platforms - How to offer eCommerce on OTT systems without having to make any changes to the platform or viewing devices?

New On and Off Screen Revenue Sources - What are new eCommerce revenues for OTT services? How to earn revenue after subscribers leave the platform (off screen)?

Affiliate Commission Programs - How to find and create relationships with vendors who are willing to pay a commission for referrals from the OTT platform? How to set up automatic ways to track and manage referral commissions?

OTT Shopping Channels - How to produce and host TV shopping channels? How to create product videos for shopping channels? What promotion services can OTT providers offer for shopping channel vendors?

OTT Order Processing - How to set up and run TV shopping carts? How to use virtual call centers? How to set up OTT eCommerce payment processing options, including adding it to TV bills, gift cards or credit cards?

tCommerce Business

The key parts of OTT Television Commerce business include affiliate referral services, direct response sales and shopping channels.

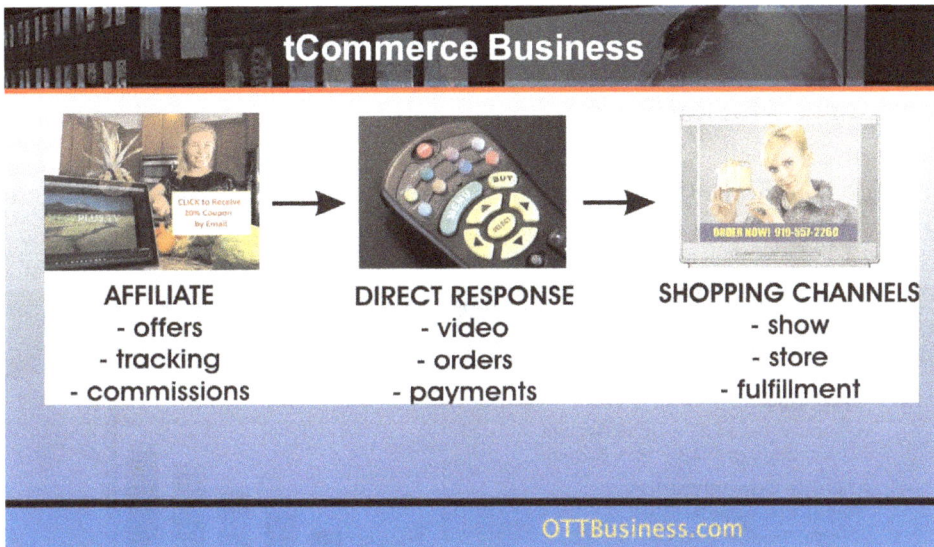

Affiliate programs involve managing offers, tracking referrals and orders and collecting commissions. Affiliate programs can be as simple as putting Amazon affiliate offers on the OTT platform websites and in promotional media posts, or including affiliate links in video ads for affiliate partners.

Direct response revenue comes from the promotion of products or services where viewers can immediately purchase products. Basically, infomercials on steroids. The OTT provider sets up video promotion shows or ads, phone numbers and order portals that come from the OTT platform. Orders and payments may be processed on the OTT platform on a partner portal.

OTT providers can earn revenue from live or pre-recorded shopping channels similar to QVC, HSN, along with video catalogs, short length infomercials and other ways to directly promote and earn revenue from promotional product videos. Show episodes typically demonstrate the use and benefits of products allowing customers to call in their orders by using online or TV shopping carts.

tCommerce Marketplace

Television commerce (tCommerce) is part of the eCommerce worldwide marketplace, the revenue of which was 4.2 trillion U.S. dollars in 2019.

eCommerce Marketplace

The eCommerce marketplace is projected to grow to over $7 trillion by 2023 [source: statista]. While most of eCommerce revenue is processed on smartphones and PCs, it is relatively easy to enable eCommerce transactions to smart TVs because they are internet connected viewing devices.

Teleshopping revenue from QVC, HSN and other shopping channels was $44B, growing at 1% each year. The growth of teleshopping revenue may be slowed by the transition of viewers from HSN, QVC and other shopping channels to online video channels such as Youtube.

Affiliate Commissions

OTT providers can earn revenue by making referrals through a clickable link that includes a tracking code, which earns the OTT provider a commission when they buy off the OTT platform.

Affiliate commissions are approximately 5% commission on the sale of physical products, such as those through Amazon's affiliate program, and can be 30% or more on digital products, such as on Commission Junction's platform - CJ.com

eCommerce Consumption

The average amount of online purchases that are made annually (eCommerce consumption) per person was around $590 in 2019 in the U.S. The UK has the highest annual online purchase amount per capita, at over $1000 in 2019.

Commerce Market Shares

Brick and mortar businesses (retails stores) made up 84% of sales in 2019, and are decreasing in rate, while online sales (eCommerce) is 16% of the marketplace and is increasing.

tCommerce Evolution

eCommerce services and applications need to be adapted so they can work on television systems. While many of the processing options on connected TV devices are similar to online store platforms, some key differences include control options, screen size, software program operation, screen viewing time and order processing.

tCommerce Evolution

- Interactive TV Trials
- QUBE – Columbus - 1977
- FSN – Orlando - 1994
- Enhanced TV – ETV - 2012
- tCommerce Apps - 2017

OTTBusiness.com

tCommerce History

tCommerce has been tried many times and has encountered several failures. Interactive TV examples include the QUBE system in Columbus Ohio in the 1970s, the FSN system in Orlando in the 1990s and the evolving enhanced TV - ETV OpenCable standards developed in 2012. These systems did generate revenue, but not enough to continue growth and expansion.

Over the Top tCommerce

Adding eCommerce to OTT systems in the 2020s does not require changes to devices, system equipment or even middleware. OTT systems can implement tCommerce in apps without any required changes to end user devices or broadcast systems.

eCommerce Shift to Television

OTT tCommerce apps provide offers that allow users to interact and make purchases. The transactions are sent through the internet to order processing and fulfillment systems that do not need to be part of the OTT system.

User Interface

User interface (UI) are the ways TV controls can be used to navigate product pages and purchase items in shopping carts. OTT tCommerce uses different user controls and devices, such as TV remotes, which have limited keypads.

Example - Viewers who login or sign up for a new account need to use a letter selection screen because the TV remote does not have a full keyboard. The selection and entry of letters using a virtual keyboard and a TV remote is very time consuming. To solve this challenge, it may be possible to use a second screen smartphone as a control option.

Screen Size

Typical TV apps are created for small to midsize screens. Running mobile or PC apps on large screen TVs looks bad. Large screens need more content, different layout and TV-like navigation options.

tCommerce Software

Multiple versions of tCommerce software programs are needed for different types of viewing devices. Televisions do not have the same software processing options as mobile and PC devices, therefore special features need to be processed in a media adapter or by the OTT system. Feature operations on multiple types of devices need to provide a similar user universal operation.

Screen Viewing Time

Consumers spend four to five times longer viewing content on big tv screens than they do on smaller smartphones. The extra viewing time on large screens provides more product placement opportunities and promotional media.

TV Order Processing

Order processing on television involves interacting with forms and displaying orders using a TV remote control. Because TV remotes have a limited number of buttons, it can be time consuming for the viewer. To solve this challenge, a TV shopping cart can be set up with easy data entry (pre-filled data) to include ways customers can get support for their tv, including on-screen and mobile app order status updates and using virtual call centers.

tCommerce Services

OTT tCommerce service options include affiliate (off-screen) referrals, direct response TV (on screen purchases), and OTT shopping channels.

tCommerce Affiliate Programs

tCommerce Affiliate Programs use the OTT system to insert affiliate product promotion offers, which refer viewers to affiliate partners with trackable links or codes.

Affiliate Referrals

Affiliate referrals motivate and enable prospective customers to go to an affiliate partner to make purchases. Referrals may be immediate or delayed, allowing commissions to be earned on and off screen.

OTT providers can put affiliate product offers on their website, in promotional media posts such as Youtube and blogs, and into show and program content.

Affiliate Cookies

Referral Tracking can be done by viewer history or coded tags (cookies) saved on the referred customers device that allow an affiliate partner to determine from where and when they were referred. Viewing and referral data can be anonymized to maintain viewer privacy.

Affiliate links are URLs with codes attached to allow the tracking of people who select them. Affiliate links can be included with any clickable link or graphic.

Affiliate Order Processing

Affiliate order processing and fulfillment is performed by the affiliate partner.Using affiliate programs, no customer support is required from the OTT service provider. This allows an OTT provider to set up a new revenue service without the need to set up or run a fulfillment system.

Affiliate Commissions

Affiliate commissions typically range from 5% to 20% of total sale. Commissions tend to be higher on digital products (e.g. ebooks, subscriptions, videos, etc) because the margins are higher and fulfillment is automatic.

Affiliate networks are platforms and marketplaces that allow vendors (affiliate promoters) to offer and track affiliate commissions to co-marketing partners (publishers).

Direct Response TV (DRTV)

Direct response TV (DRTV) is the providing of television content (such as an ad or infomercial) where the viewer can directly respond to the promotional media by on screen, website, phone or other real time response process.

Direct Television Promotion

OTT direct response TV is the insertion of promotional content into broadcast or streaming services that includes offers with immediate response options.

Direct Response Order Processing

Direct response order processing enables viewers to immediately place an order or get additional information. Order processing services may be completed by advertisers, fulfillment services, or other companies that receive and process orders during the TV showing.

Direct Response TV Show Services

DRTV Services can include infomercial show streaming, transaction processing, product fulfillment, customer support and show production.

OTT Shopping Channels

OTT shopping channels include product promotional videos, host promotion and viewer engagement.

OTT shopping channels typically include product video clips, host demonstrations and order processing services.

Live TV shopping channel productions can be created in studios or remote production sites that allow for the demonstration of products and immediate communication with customers.

Pre-Recorded TV Shopping Channel Productions use pre-produced product video content. Production fees may be paid for by the product vendor. Pre-recorded TV shopping channels are facilitated by a moderator and can include live audience media, such as questions, comments, and photos from audience participants. A key advantage for pre-recorded shopping channels is the ability to review audience engagement and optimize (edit, enhance) program content.

Key tCommerce Companies & Systems

Key tCommerce company types include affiliate networks, remote production platforms and OTT shopping carts.

TV eCommerce Affiliate Networks

TV eCommerce affiliate networks and services manage and pay commissions on co-promotion marketing programs between companies and publishers. Key TV eCommerce affiliate networks include Amazon, Commission Junction, Rakuten and others.

Live Remote TV Production

Live remote TV production companies or platforms are used to produce live streaming events and shows. Key live remote production companies include Livestream, Twitch.tv/OBS, DaCast and WireCast.

OTT TV Shopping Carts

OTT TV shopping carts are systems or services that can display product offers (video catalogs) and process orders through television platforms. Key OTT TV shopping cart providers include Roku, Muvi and TalkShop.live.

For an expanded list of tCommerce Companies, visit the OTT Business Opportunities Student Resource Area.

Appendix 1 - Terms & Acronyms

AAC-Advanced Audio Codec
ABR-Adaptive Bit Rate
Ad Agency-Advertising Agency
Ad Campaign-Advertising Campaign
Ad Insert-AD Insertion
Ad Splicer-Advertising Splicer
ADC-Analog to Digital Conversion
ADS-Ad Decision Service
Affiliate-Network Affiliate
AFP-Advertiser Funded Program
AI-Artificial Intelligence
ATSC-Advanced Television Systems Committee
Avail-Ad Availability Slot
Avails-Insertion Opportunities
AVOD-Advertising Video on Demand
Broadcast TV-Broadcast Television
Calls-Casting Call
CAS-Conditional Access System
CATV-Cable Television
CATV-Community Access Television
CDN-Content Delivery Network
CMS-Content Management System
Codec-Coder/Decoder
Connected TV-Connected Television
CP-Content Protection
CPA-Cost Per Action
CPI-Cost Per Install
CPM-Cost Per Thousand
CSAI-Client Side Ad Insertion
CTR-Click Through Rate

DAI-Dynamic Ad Insertion
DMA-Designated Market Area
DMA-Digital Media Adapter
DMP-Data Management Platform
Dongles-TV Media Stick
DRM-Digital Rights Management
DRTV-Direct Response Television
DSP-Demand Side Platform
DTT-Digital Terrestrial Television
DTV-Digital Television
eCare-Electronic Customer Care
eCommerce-Electronic Commerce
eCPM-Effective Cost Per Thousand
EPG-Electronic Program Guide
ES-Elementary Stream
FLV-Flash Video
for Distribution-Program Packaging
Freeview-Free View
Geocoding-Geographic Coding
Geotargeting-Geographic Targeting
Gridcasting-Peercasting
HbbTV-Hybrid Broadcast Broadband Television
HDTV-High Definition Television
HMN-Home Media Network
HMS-Home Media Server
Hosted Ads-Hosted Advertisements
HSTB-Hybrid Set Top Box
HTML5-Hypertext Markup Language Version 5
iDTV-Interactive Digital Television
Indies-Independent Programming
Internet TV-Internet Television
IP Broadcast-Internet Protocol Broadcast
IP STB-Internet Protocol Set Top Box
IPG-Interactive Programming Guide
IPTV-Internet Protocol Television
ISO-International Standards Organization

ISTB-Internet Set Top Box
I-STB-Internet TV Set Top Box
LBA-Location Based Advertising
Linear TV-Linear Television
MG-Media Gateway
MJPEG-Motion JPEG
ML-Machine Learning
MP1-Moving Picture Experts Group Layer 1
MP2-Moving Picture Experts Group Layer 2
MP3-Moving Picture Experts Group Layer 3
MP3Pro-MPEG Layer 3 Pro
MP4-MPEG-4
MPEG-Moving Picture Experts Group
NTSC-National Television System Committee
NVOD-Near Video on Demand
OEM-Original Equipment Manufacturer
OpApp-Operator Application
OS-Operating System
OSP-OTT Service Provider
OTT-Over the Top Television
Pause Ads-Pause Advertising
PCR-Production Control Room
PCs-Multimedia Personal Computers
PPP-Pay per Period
PPV-Pay Per View
PR-Public Relations
PR Agent-Publicity Agent
Preview Ads-Preview Advertisements
Props-Scenic Elements
PVO-Promotional Video Optimization
PVOD-Premium Video on Demand
RTP-Real Time Transport Protocol
SaaS-Software as a Service
Script-Screenplay
Scripting-Script Writing
SDK-Software Development Kit

SDV-Switched Digital Video
SFX-Special Effects
SMPTE-Society of Motion Picture and Television Engineers
Spots-Advertising Spots
SPTS-Single Program Transport Stream
Squeezeback-Squeeze Back
SSAI-Server Side Ad Insertion
SSP-Supply Side Platform
STB-Set Top Box
Studio-Movie Studio
Sublicensing-Sub-Licensing
SVOD-Subscription Video on Demand
SVS-Switched Video Service
tCommerce-Television Commerce
Temporal Compression-Time Compression
Terrestrial TV-Terrestrial Television
Time Compression-Temporal Compression
Timestamp-Time Stamp
TVOD-Transaction Video on Demand
TS-Transport Stream
TV Broadcaster-Television Broadcaster
TV Channel-Television Channel
TV Server-Television Server
TV Station-Television Station
TV Studio-Television Studio
TVoDSL-Television over DSL
UGC-User Generated Content
UI-User Interface
UX-User Experience
VAST-Video Ad Serving Template
VC-1-Windows Media
Video Player Speed Control-Trick Mode
VOD-Video On Demand
Webinar-Web Seminar
WFM-Workflow Management

Appendix 2 - OTT Directory

OTT Business Directory
Companies * Services * Tools

This is a list of Companies, Services, and Tools that come with OTT Business Opportunities Course & Book. These resources help producers, distributors, and promoters to create, run, and monetize streaming TV services and systems.

OTT Systems and Services
Companies that Provide Streaming TV Services

OTT Service Providers-Companies that provide movie and TV show video streaming services.

OTT Platforms-Systems and services that list, schedule, and stream TV shows and programs.

Live Streaming Systems-Hardware and software systems that are used to produce live streaming events and shows.

TV Captioning-Systems and services that transcribe and add audio captioning to movies, TV shows, and videos.

Digital Asset Management (DAM)-Systems and services that store, organize, and manage movies, TV shows, and video content.

Cloud Video Production-Services that enable the selection, editing, and publishing of movies, TV shows, and video through the Internet.

Content Distribution Networks (CDNs)-Services that manage distribution of movies, shows, and videos to multiple points.

OTT Advertising
Companies that Provide Ad Systems & Services

Video Ad Networks-Are platforms or services that enable advertisers to upload ready to publish video ads and for online TV systems to select and insert the ads into their program content.

Advertising Measurement Services-Companies or platforms that measure ad viewing and engagement activities.

TV Apps
Companies that Develop & Distribute TV Apps

TV App Developers-Software development companies that specialize in creating video apps for Smart TVs, digital media adapters, and other connected TV devices

TV App Distributors-Companies or platforms that submit and transfer apps to app multiple marketplaces, systems, and devices.

TV App Marketplaces-Online web portals where users can find and download TV and mobile apps for OTT services.

Social TV
Companies that Provide Social Media & Clipping Platforms & Services

Social TV Platforms-Systems and services that gather, organize, and publish social media related to OTT channels and services.

TV Video Clipping Services-Systems or platforms that can find, select, describe, package, and publish movie, TV show, OTT channel related video clips.

Television Commerce (tCommerce)
Companies that Provide TV eCommerce Platforms & Services

TV eCommerce Affiliate Platforms-Systems and services that manage and pay commissions on co-promotion marketing programs between companies and publishers.

TV Shopping Carts-Systems and services that enable OTT viewers to view, select, and purchase products on their televisions using TV shopping carts.

Hybrid TV Systems-Television systems that combine broadcast television and broadband Internet TV (OTT) services.

50+ More Categories.......

500+ Companies, Services, & Tools...

To get access to these materials, go to http://OTTBusiness.com/ottdirectory

OTT Magazines

OTT Magazines publish articles, news, and information on how to select, setup and run Streaming TV systems, services, and promotion activities.

OTT Business Magazine - OTTBusiness.com/magazine
OTT Exec - OTTExec.com/magazine
Streaming Media - StreamingMedia.com/magazine

To get an expanded list of OTT Magazines, go to OTTBusiness.com/ottmagazines

OTT Resource Platforms

OTT resource platforms share sample plans, guides, templates, contracts, and other materials that are used to setup, run, and promote Streaming TV systems & services.

OTT Business Resources - OTTBusiness.com/resources
No Film School - https://nofilmschool.com/2016/08/grab-every-filmmaking-form-youll-ever-need-these-99-free-templates

To get an expanded list of OTT Resource Platforms, go to OTTBusiness.com/ottresources

OTT Podcasts

OTT podcasts are audio or video shows distributed online that cover ways to select, setup, run, and promote Streaming TV systems & services.

OTT Business Podcast - OTTBusiness.com/podcast
TV of Tomorrow Televisionation - thetvoftomorrowshow.com/televisionation
Muvi - muvi.com/podcast.html

To get an expanded list of OTT Resource Platforms, go to OTTBusiness.com/ottpodcasts

OTT Research Companies

OTT research companies gather, organize, and present Streaming TV market trends and consumer behaviors.

Digital TV Research - digitaltvresearch.com
Grandview Research - grandviewresearch.com
Parks Associates - parksassociates.com

To get an expanded list of OTT Research Companies, go to OTTBusiness.com/ottresearch

OTT Courses

OTT courses explain OTT systems and services options, how they work, and ways to run and optimize them.

OTT Business Opportunities - OTTBusiness.com/ottcourse
IABM Courses - theiabm.org/technical-training/
OTT TV Training - Tonex - www.tonex.com/ott-tv-training-over-the-top-tv/

To get an expanded list of OTT Courses, go to OTTBusiness.com/ottcourses

OTT Books

OTT books explain OTT systems and services options, how they work, and ways to run and optimize them.

OTT Business Opportunities - OTTBusiness.com/ottbook
Internet TV Systems - https://www.amazon.com/dp/1932813268
Android TV Apps Development - https://www.amazon.com/dp/1484217837/

To get an expanded list of OTT Books, go to OTTBusiness.com/ottbooks

OTT Discussion Groups

OTT discussion groups share insights, questions, and resources that help to select, setup, and run OTT and Streaming TV systems.

OTT Business - linkedin.com/groups/12484553
Internet TV Systems - linkedin.com/groups/2863304
Android TV - reddit.com/r/AndroidTV/

To get an expanded list of OTT Discussion Groups, go to OTTBusiness.com/ottgroups

OTT Events

OTT events are conferences (live or online) that enable companies to share product and service information to attendees.

OTT and Video Distribution Summit - ottvideodistributionsummit.com
OTT Exec - OTTExec.com
NAB - nabshow.com
IBC Show - show.ibc.org

To get an expanded list of OTT Events, go to OTTBusiness.com/ottevents

OTT Provider & Platform Companies

Key OTT company types include streaming TV service providers, platform providers, access device manufacturers and security systems.

OTT Service Providers (OSPs)

OTT service providers (OSPs) supply streaming programs directly to consumers. There are thousands of OSPs, and there is a growing trend of successful niche content providers.

Netflix - Netflix.com
Amazon Video - Amazon.com/amazonprime
Hulu - Hulu.com
Curiosity Stream - CuriosityStream.com

To get an expanded list of OTT Service Provider (OSP) Companies, go to OTTBusiness.com/ottserviceproviders

OTT Platform Providers

OTT platform providers set up systems and software that store and stream video content for OSPs.

Brightcove - Brightcove.com
Kaltura - Kaltura.com
Dacast - Dacast.com

To get an expanded list of OTT Platform Companies, go to OTTBusiness.com/ottplatforms

Hybrid TV Systems

Hybrid TV systems combine broadcast television and broadband TV (OTT) services.

HbbTV - hbbtv.org
ATSC 3.0 - atsc.org/nextgen-tv
Hybridcast - iptvforum.jp

To get an expanded list of Hybrid TV Systems, go to OTTBusiness.com/otthybrid

Access Device Manufacturers

Access device manufacturers or their original equipment manufacturer (OEM) who produce media players and adapters.

Roku - roku.com/products/players
Google Chromecast - google.com/chromecast
Amazon Firestick - amazon.com/firestick

To get an expanded list of OTT Access Devices, go to OTTBusiness.com/ottdevices

OTT Security System Companies

OTT security system companies develop software systems and key software, privacy and content protection services.

Pallycon (Widevine) - pallycon.com
Verimatrix - verimatrix.com
Irdeto - irdeto.com

To get an expanded list of OTT security system companies, go to OTTBusiness.com/ottsecurity

OTT Advertising Companies

The key types of companies that set up and run advertising services for OTT providers include ad networks, server side platforms, data management platforms, demand side platforms, ad measurement services and dynamic video services.

OTT Advertising Management Platforms

OTT advertising management platforms identify and connect advertising insertions opportunities (ad inventory) with advertiser campaigns (ad offers).

Decentrix - decentrix.com
Strategus - strategus.com
Lightcast - lightcast.com/ott-advertising/
xandr - xandr.com/platform/

To get an expanded list of OTT advertising management platform companies, go to OTTBusiness.com/ottadmanagement

Video Ad Networks

Video ad networks are marketplaces that link advertisers to publishers and allow for the automatic requesting, selection and insertion of ads.

Google DFP - google.com/ads/publisher/
Media.net - media.net
Auditude - adobe.com/products/auditude.html

To get an expanded list of OTT ad networks, go to OTTBusiness.com/ottadnetworks

Ad Supply Side Platforms (SSPs)

Ad Supply Side Platforms (SSP) are systems and services that manage and request ads.

Google Ad Manager - admanager.google.com
Open X SSP - openx.com
Pubmatic - pubmatic.com

To get an expanded list of Ad supply side platforms - SSPs, go to OTTBusiness.com/ottssps

Data Management Platforms (DMPs)

Data Management Platforms (DMPs) gather, organize and use customer data for ad targeting and delivery.

Lotame - lotame.com
Oracle DMP (BlueKai) - oracle.com/data-cloud/
OnAudience.com - onaudience.com

To get an expanded list of Ad data management platforms - DMPs, go to OTTBusiness.com/ottdmps

Demand Side Platforms (DSPs)

Demand Side Platforms (DSPs) are systems and services that provide ads, video, and other media formats, to publishers.

TubeMogul (Adobe) - tubemogul.com
Google Doubleclick - marketingplatform.google.com
Amazon Advertising - advertising.amazon.com

To get an expanded list of Ad demand side platforms - DMPs, go to OTTBusiness.com/ottdsps

Advertising Measurement Services

Advertising measurement service providers are companies and platforms that measure ad viewing and engagement.

Nielsen - nielsen.com
Moat Analytics - moat.com
Conviva - conviva.com
C3 Metrics - c3metrics.com

To get an expanded list of advertising measurement services, go to OTTBusiness.com/ottadmeasurements

Dynamic Video Services

Dynamic video service provider companies provide advanced video analytics and real time video rendering services.

Iris.tv - iris.tv
xandr - xandr.com
Innovid - innovid.com

To get an expanded list of dynamic video service companies, go to OTTBusiness.com/ottdynamicvideoads

TV App Companies

The key types of companies involved in TV apps include TV app developers, app marketplaces and app distributors.

TV App Developers

TV app developers are software development companies that specialize in creating video apps for Smart TVs, digital media adapters and other connected TV devices.

Accedo - accedo.tv
24i - 24i.com
Applicaster - applicaster.com

To get an expanded list of TV app development companies, go to OTTBusiness.com/ottappdevelopment

TV App Marketplaces

TV app marketplaces are online portals where users can find and download TV apps.

Android TV - android.com/tv
Apple TV - apple.com/apple-tv-app
Vewd (Opera) - vewd.com/products-services/vewd-app-store/
Samsung - samsung.com/my/tvs/smart-tv-apps/
Aptoide TV - https://tv.aptoide.com/

To get an expanded list of TV app marketplaces, go to OTTBusiness.com/ottappmarketplaces

TV App Distributors

TV app distributors are companies that submit apps to multiple marketplaces.

Zype - zype.com
Muvi - muvi.com
MobiTV - mobitv.com

To get an expanded list of TV app distributors, go to OTTBusiness.com/ottappdistributors

Social TV Companies

Key Social TV company types include social TV platforms, social TV ratings, fan management and TV video clipping services.

Social TV Platforms

Social TV platforms gather, organize and publish social media related to OTT channels and services.

Never.no - never.no
Flowics - flowics.com
Vizrt - vizrt.com
Actus Digital - actusdigital.com
LTN Global (Make TV) - ltnglobal.com

To get an expanded list of social TV platforms, go to OTTBusiness.com/ottsocialtv

Social TV Ratings

Social TV ratings companies review, analyze and measure TV viewer media feedback engagement on social media.

Nielsen Social - nielsensocial.com
Kantar - kantarmedia.com
Share Rocket - sharerocket.com/

To get an expanded list of social TV ratings companies, go to OTTBusiness.com/ottsocialratings

Fan Management

Fan management companies and platforms identify, store, categorize, manage, send messages and track fan activities.

Monterosa (Lvis) - monterosa.co
Fantrust - fantrust.com
Telescope TV - telescope.tv

To get an expanded list of fan management platform companies, go to OTTBusiness.com/ottfanmanagement

TV Video Clipping

TV video clipping companies and services find, select, describe, package and publish TV video clips.

Graybo - graybo.com
Blackbird - blackbird.video
NewTek - newtek.com

To get an expanded list of OTT video clipping platform companies, go to OTTBusiness.com/ottvideoclipping

tCommerce Companies

Key tCommerce company types include affiliate networks, remote production platforms and OTT shopping carts.

TV eCommerce Affiliate Networks

TV eCommerce affiliate networks and services manage and pay commissions on co-promotion marketing programs between companies and publishers.

Amazon - affiliate-program.amazon.com
Commission Junction - cj.com
Rakuten - rakuten.com

To get an expanded list of affiliate platform companies, go to OTTBusiness.com/ottaffiliate

Live Remote TV Production

Live remote TV production companies or platforms are used to produce live streaming events and shows.

Livestream - livestream.com
Twitch.tv - twitch.tv
DaCast - dacast.com
WireCast (Telestream) - telestream.net/wirecast

To get an expanded list of live stream platform companies, go to OTTBusiness.com/ottlivestream

OTT TV Shopping Carts

OTT TV shopping carts are systems or services that can display product offers (video catalogs) and process orders through television platforms.

Roku - roku.com
Muvi - muvi.com/muvi-kart.html
TalkShop.live - talkshop.live

To get an expanded list of TV shopping cart platform companies, go to OTTBusiness.com/ottshoppingcarts

Appendix 3 - OTT Resources

OTT Business Resources
*Guides * Plans * Templates*

This is a list of Resources that come with OTT Business Opportunities Course & Book. These resources help producers, distributors, and promoters to create, run, and monetize streaming TV services and systems.

OTT Service Business Plan-Key objectives, service descriptions, audiences, facilities, systems, production, key activities, resources.

OTT Media Channels-Spreadsheet that lists 20+ typical promotion channels (Twitter, Blog, FaceOTT, Pinterest, Quora, etc), name, URL, purpose, strategies, & recommended posting schedule.

OTT Glossary-Terms acronym definitions used in the OTT and streaming TV industry.

OTT Business Procedures-Sample steps and processes on how to produce, distribute, and promote streaming TV services.

OTT Viewer Interview Guide-How to find and talk to streaming TV viewers to discover key topics, activities, & media channels for more effective promotion.

OTT Directory-Lists of companies, services, and tools for streaming TV production, distribution, and promotion.

OTT Channel Distribution Agreement Template-Sample agreement for licensing and providing a channel to streaming TV and VOD platforms.

OTT Channel Product Placement Agreement-Sample agreement for that allows a promoter to pay for content insertion into a channel or media media platform.

OTT TV App Development Agreement-Sample agreement that defines the features, platform, and services for the creation, distribution, and operation of TV apps.

OTT TV App Distribution Agreement-Sample agreement that manages the distribution, installation, and updating of apps on devices and platforms.

Contribution and Use Permission Forms-Sample agreement documents that provide written verification to use contributed photos and materials in your shows and platforms.

OTT Media Agent Agreement-Sample agreement for contractors or freelancers who create and publish media posts about your show.

OTT Advertising Rate Sheet Template-Example rate sheet for OTT advertising services.

OTT Advertising Agreement-Sample agreement for OTT advertising services.

OTT Social Media Rules-Sample rules for OTT social media publishing rules and policies.

Social TV Production Agreement-Sample agreement for companies or contractors to find, process, and publish clips and media about OTT channels & shows.

Affiliate Commission Agreement Template-Example co-marketing agreement between OTT service providers and vendors who pay commissions on referred product sales.

OTT Shopping Channel Agreement-Sample agreement for OTT service provider to host, produce, and/or promote a TV shopping channel for a vendor.

100+ More.......

To get access to these materials, go to http://OTTBusiness.com/ottresources

Index

www.ingramcontent.com/pod-product-compliance
Lightning Source LLC
Chambersburg PA
CBHW080558220326
41599CB00032B/6528